A FUNCTIONAL BIOLOGY OF NEMATODES

A Functional Biology of Nematodes

David A. Wharton

The Johns Hopkins University Press
Baltimore, Maryland

 Published 1986
The Johns Hopkins University Press,
701 West 40th Street, Baltimore, Maryland 21211
Printed and bound in Great Britain

Library of Congress Cataloging in Publication Data

Wharton, David A.
A functional biology of nematodes.

Bibliography: p.
Includes index.
1. Nematoda. I. Title.
QL391.N4W46 1986 595.1'82 86-7213
ISBN 0-8018-3359-0

CONTENTS

For Ann

PREFACE

I am continually surprised and delighted by the variety and beauty of the adaptations that enable nematodes to exploit their various habitats. Sometimes these are almost unbelievable. Some nematodes can remain active for several hours in fixative, and can survive exposure to extreme desiccation and vacuum, and the eggs of some species can survive exposure to concentrated sulphuric acid. The ability to survive periods of exposure to adverse environmental conditions has enabled nematodes to exploit almost every conceivable habitat. For activity, growth and reproduction to occur, however, at least a film of water must be present.

The economic importance of parasitic nematodes has focused attention on these, rather than free-living species. Despite the abundance of free-living nematodes, they have been relatively little studied, and this book, of necessity, reflects this imbalance. It has, however, been my intention to examine the biology of the phylum as a whole and, where possible, examples have been drawn from animal parasitic, plant parasitic and free-living species. The free-living nematode *Caenorhabditis elegans* has in recent years been adopted as a convenient model to study important problems in the developmental genetics and biology of metazoans. Clearly, our knowledge of the biology of free-living nematodes and of nematode biology as a whole will increasingly be based on this species.

Any author attempting to write on a group as widespread as the Nematoda is in danger of being overwhelmed by the volume of information which he unearths. My approach has been to try to provide an overview of the features of the biology of nematodes which seem to me to be important. Above all I have attempted to show how nematodes work as integrated wholes, and how they are able to live in the habitats they do. I have not tried to give detailed descriptions of nematode morphology and the structural variations between species, but have tried to provide references where such details can be found. I hope that enough morphological detail has been provided to enable the functional organisation of nematodes to be understood.

Many people have helped and encouraged me in the writing of the book. Thanks are due especially to Professor John Barrett, without whose aid and advice over the past few years I would never have been in a position to undertake such a task. Thanks also to John Barrett, Stephen Young, Ann Wharton, Rolo Perry, Tim Hardwick and Peter Calow who read the entire manuscript, and to Bob Wootton, Adrian Hopkins, Lloyd Davis, Mike Barker, Rod Turner, Marcus Trett and Chris Preston who read parts of it. Julia Bowler drew or completed the drawing of most of the figures and Jean Clough drew some of the figures in Chapter 6. The following kindly provided me with copies of published figures and/or plates: Dr D.J. McLaren, Dr A.M. Shepherd, Dr C.C. Doncaster, Dr R.I. Sommerville, Dr J. Giebel, Dr M.G.K. Jones, Dr T. Jenkins, Dr H.C. Bennet-Clark, Dr E.A. Munn, Professor K.A. Wright, Dr P.A.G. Wilson, Professor H.R. Wallace, Dr J.G. White and Dr C.M. Preston. The sources of redrawn figures and published data are acknowledged in the text. I would also like to thank the following for permission to reproduce material: Academic Press, Macmillan Publishers, Cambridge University Press, Pergamon Press, Annual Reviews Inc., Longman Group Ltd, *Annals of Applied Biology*, Zoological Society of London, Society of Nematologists, Elsevier/North Holland Press, and the Royal Society of London. The sources of published figures and plates are given in the text.

David A. Wharton

1 INTRODUCTION

Within the animal kingdom, nematodes are second only to the arthropods both in the numbers of species and in the numbers of individuals present. Indeed, they are so numerous that, in a famous quote, Cobb (1915) was led to comment:

> If all the matter in the universe except the nematodes were swept away, our world would still be dimly recognisable, and if, as disembodied spirits, we could investigate it, we should find its mountains, hills, vales, rivers, lakes and oceans represented by a thin film of nematodes ...

Although they have a relatively constant body form, nematodes have successfully exploited a wide variety of habitats. They are parasites of almost every species of animal and plant. Free-living species live in the soil, and in freshwater and marine habitats. They may be bacterial-feeders, fungal-feeders, algal-feeders, herbivores, omnivores, parasites or predators. They are found in hot springs, were thawed out from the Antarctic ice by members of the Shackleton expedition, and can survive complete desiccation and exposure to high vacuum. Their only limiting factor appears to be that, although some species can survive desiccation, activity is limited to environments where they are surrounded by at least a film of water.

Approximately 48 per cent of nematode genera are parasites of animals and plants (Table 1.1). The remainder are free-living in the soil and in freshwater or marine sands and muds. These figures are probably heavily biased towards parasitic nematodes. The large number of genera of vertebrate parasites reflects the amount of study that has been devoted to this group as well as the richness of the fauna. Around 1500 to 1600 nematode species have been described, and it has been estimated that there are about 42000 species in total (Poinar, 1983). Free-living nematodes have been relatively little studied, and it is clear that many species remain to be discovered. Parasitic species may, therefore, be heavily outnumbered by free-living species.

Despite the ubiquity of nematodes they rarely reach the con-

Table 1.1: Approximate Numbers of Nematodes in Various Habitat Classes

Habitat	Families	Genera	Total %
Marine and freshwater	41	730	33
Soil	64	429	19
Plant parasites	26	166	7
Invertebrate parasites	42	187	8
Vertebrate parasites	83	759	33
Totals	256	2271	100

From Anderson (1984)

sciousness of the general public. Even among zoologists the study of nematodes has traditionally been regarded as the preserve of the 'parasitologist' or 'applied biologist' rather than worthy of the attention of invertebrate physiologists. The success of nematodes as parasites is perhaps responsible for this attitude. Nematodes cause many economically important diseases of humans, domestic stock and plant crops. Their study has largely been undertaken by the medical and veterinary helminthologist and the plant nematologist. This has also had the effect of dividing those who work on animal parasitic nematodes ('parasitologists' or 'helminthologists') from those who work on plant parasitic nematodes ('nematologists'). They have different professional associations, publish in different journals, and may even use different terminologies. Free-living nematodes have not received the attention that their abundance in many habitats suggests that they deserve.

The study of nematodes as parasites has focused attention on the pathology, epidemiology and control of nematode diseases rather than upon the biology, physiology and ecology of the nematodes themselves. To some extent the balance has been redressed in recent years by the use of nematodes as model systems to study fundamental biological problems (Zuckerman, 1980). *Caenorhabditis elegans*, a free-living nematode, is easy to culture, has a short generation time and possesses a number of mutants suitable for genetic analysis (Brenner, 1974). It is rapidly becoming the metazoan equivalent of *Escherichia coli.* A considerable amount of effort continues to be expended on the study of this animal, and it is clear that much of our detailed knowledge of nematode biology will be based upon this. The dangers of a 'model system' such as this are that nematode biology may be forgotten in the quest for fundamental biological principles and that our knowledge of nematodes may largely be based on a single convenient, but perhaps atypical, species; as has been the case with *Ascaris lumbricoides* in the past.

This book aims to present what is known about the structure and physiology of nematodes in the context of invertebrate biology and ecological physiology. It is aimed at the research worker and the final-year undergraduate. For introductory texts the reader should refer to Crofton (1966), Croll and Matthews (1977), Poinar (1983) and Maggenti (1981). Information on animal parasitic nematodes is given by most parasitology texts (e.g. Cheng, 1973), on plant parasitic nematodes by Wallace (1963) and Norton (1978) and on free-living nematodes by Nicholas (1984). There are more advanced texts by Bird (1971) on the structure of nematodes, Lee and Atkinson (1976) on nematode physiology, and the classic *Introduction to Nematology* by Chitwood and Chitwood (1974).

General Organisation

The body of a typical nematode consists of a flexible cylinder which tapers at both ends, with a pointed tail and a blunt head (Figure 1.1). The body is bounded by a flexible but tough collagenous cuticle. Beneath the cuticle is a hypodermis (= epidermis) which may be cellular or syncytial. A layer of longitudinal muscles

Figure 1.1: Diagrammatic Representation of the Morphology of a Nematode (Female), A, anus; AD, anal dilator muscles; AM, amphid; BC, buccal capsule; C, cuticle; CP, cephalic papilla; EP, excretory pore; ES, excretory system (median duct); I, intestine; NR, nerve ring; OE, oesophagus; OV, ovary; P, phasmid; PC, pseudocoelomocyte; PS, pseudocoel; SR, seminal receptacle; T, tail; UT, uterus; V, vagina. Original, not to scale

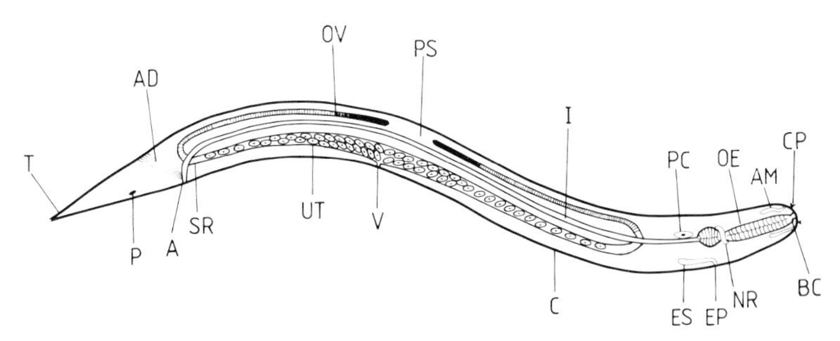

underlies the hypodermis. The hypodermis is expanded in four places to form the dorsal, ventral and two lateral hypodermal cords. These divide the muscle cells into four fields. In most species the cuticle overlying the hypodermal cords is expanded to form the lateral alae (Figure 1.2). The definitions of dorsal, ventral and lateral in nematodes are matters of convention. In fact, many species move with the lateral alae in contact with the substrate and may, therefore, be said to lie on their sides.

The organ systems (reproductive, excretory, alimentary) are tubular, relatively simple histologically and lie mainly free within a body cavity, the pseudocoel. The nervous system consists of a nerve ring at the base of the oesophagus with sensory and motor neurons extending forwards to the head and longitudinal nerve tracts extending posteriorly within the hypodermis, which connect with a posterior nerve ring or commissure and the sensory neurons of the posterior sense organs (see Chapter 3). There are no specialised circulatory or respiratory systems.

Nematodes are thus triplobastic (three layers of cells: ectoderm, mesoderm and endoderm) and possess a body cavity, the pseudocoel. They have distinct, although simple, organ systems

Figure 1.2: Generalised Cross-section of a Nematode (Female) at the Level of the Intestine. AN, annulation; C, cuticle; DHC, dorsal hypodermal cord; DNC, dorsal nerve cord; E, egg; HY, hypodermis; I, intestine; IL, intestinal lumen; LA, lateral alae; LED, lateral excretory duct; LHC, lateral hypodermal cord; MC, muscle cell; MV, microvilli; OV, ovary; PS, pseudocoel; UT, uterus; VHC, ventral hypodermal cord; VNC, ventral nerve cord. Original, not to scale

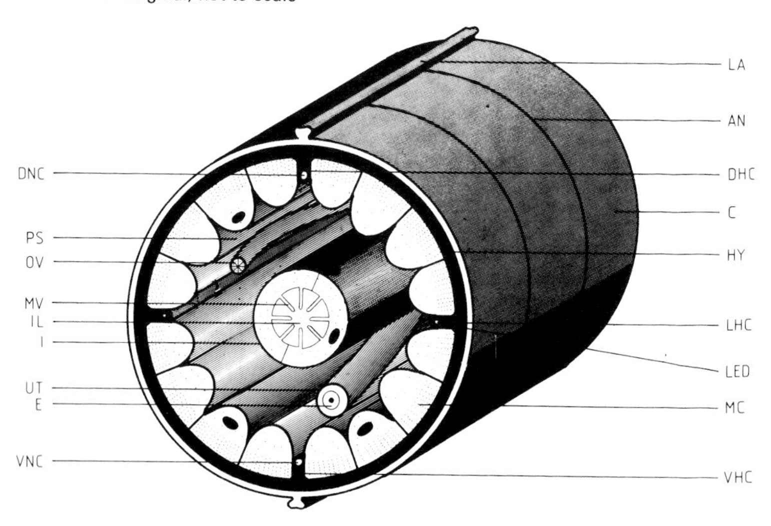

and a straight-through alimentary canal with a terminal mouth and anus. They show a basic bilateral symmetry, although there are elements of radial symmetry around the mouth. Nearly all nematodes possess this basic body plan and are very similar in appearance whether they are free-living or parasitic.

Most species have separate sexes (amphimictic), the males are smaller than the females and in many species possess accessory copulatory structures (copulatory bursae and spicules). It is usually considered that there are six stages in the life cycle: the egg or embryo, four juvenile stages and the adult. The overall body plan of the juvenile is often similar to that of the sexually mature adult, although there are often differences in the detailed structure; including, of course, that of the reproductive system. With the exception of the egg to 1st-stage juvenile, each instar is separated by the moulting of the cuticle (see Chapter 6).

Nematode Phylogeny

Nematodes are soft-bodied and have left very few fossil remains. The earliest fossils which have been attributed to a nematode are wave-like tracks found in Eocene strata (Moussa, 1969). More authenticated fossils are of nematodes preserved in amber, from mammals frozen in permafrost and from fossilised shark muscle (Conway Morris, 1981). The fossil record is too sparse to tell us much about nematode origins, and speculations about nematode phylogeny have been largely based on observations of living species.

It is difficult to determine whether similar structures in different animals are homologous, indicating a common evolutionary ancestry, or are analogous structures, which represent a similar solution to the same physiological or ecological problem. It is dangerous to base conclusions about phylogenetic relationships on comparative anatomy alone. The embryonic development of structures must be taken into account. There has been considerable disagreement regarding the phylogenetic position of nematodes and they have been placed in the Aschelminthes, the Pseudocoelomata or have been given a phylum of their own, the Nematoda or Nemata. The rationale for including the Nematoda in the phylum Pseudocoelomata along with the Rotifera, Gastrotricha, Kinorhyncha, Priapulida, Nematomorpha, Acanthocephala and Entoprocta was based on the presence of a pseudocoel. The pseudocoel was considered to be derived from the blastocoel of the developing embryo. In fact the pseudocoel appears to have different origins in the different groups of 'pseudocoelomates' and even between different species of nematodes. A body cavity is important in allowing an increase in size and complexity and for developing an efficient method of locomotion. It appears to have evolved several times, and the presence of a pseudocoel among these groups of animals is an adaptive feature and does not necessarily imply a phylogenetic relationship (Maggenti, 1976). Other features which were considered to be common among the 'pseudocoelomates', a protonephridial excretory system and eutely (constancy of cell numbers after completion of embryonic development), are not now thought to apply strictly to nematodes. It is therefore appropriate to consider them to be a separate phylum, as first proposed by Chitwood and Chitwood (1974).

If the phylogenetic position of nematodes is in doubt, their evolutionary origins have been subject to even greater speculation. It has been suggested that they are ancestors of or have descended from almost every other group of animals. They have also been proposed as ancestors for, or to have a common ancestry with, groups as diverse as the echinoderms and the arthropods (Chitwood and Chitwood, 1974). Poinar (1983) considers that the most plausible suggestion is that the nematodes arose from the gastrotrichs during the Pre-Cambrian or Cambrian period. This suggested relationship is based on morphological similarities between the two groups, particularly in the structure of the digestive and reproductive systems. Inglis (1985), however, considers that this is unlikely given differences in the structure and ontogeny of the cuticle. He suggests that the various phyla that possess a pseudocoel have separate origins from ancestral turbellarians, and are thus not part of the same phylogenetic line. Structural similarities in the pseudocoel, musculature, cuticle, anus and a triradiate oesophagus represent analogous solutions to problems of mechanical efficiency involved in an increase in size and complexity beyond that of the acoelomate condition.

Classification

The classification of nematodes is in a state of flux, with revisions frequently appearing in the literature. The classification schemes proposed naturally reflect the interests of each author, whether in free-living, plant parasitic or animal parasitic nematodes. It is usually agreed, however, that the Nematoda is divided into two classes: the Adenophorea and the Secernentia. The members of these two classes differ in a number of ways, including the presence or absence of phasmids (posterior sense organs), the form of the ovary and testis (see Chapter 4), the organisation of the excretory system and the presence of a syncytial or a cellular hypodermis. Andrassy (1976) replaces the Adenophorea with two classes, the Torquentia and the Penetrantia. Inglis (1983a) divides the Nematoda into three classes, the Enoplea and the Chromodorea (≡ Adenophorea) and the Rhabditea (≡ Secernentia).

The classification given in Table 1.2 is drawn from several sources: principally Inglis (1983a), Anderson, Chabaud and Willmott (1974-83), and Poinar (1983). It is interesting to note the distribution of habitat types among the nematode orders. Most

Table 1.2: An Outline Classification of Nematodes

Class	Sub-class	Order	Fresh water/ soils	Marine	Plant parasites	Invertebrate parasites	Vertebrate parasites
Adenophorea	Chromadoria	Araeolamida	+	+			
		Monohysterida		+			
		Desmodorina		+			
		Chromadorina		+			
		Desmoscolecida		+			
	Enoplia	Enoplida		+			
		Dorylaimida	+		+		+
		Mononchida	+				
		Mermithida	+	+		+	
Secernentia	Rhabditia	Rhabditida	+			+	+
		Oxyurida				+	+
		Strongylida					+
	Diplogasteria	Diplogasterida	+			+	+
		Drilonematida				+	
		Ascaridida				+	+
		Spirurida					+
	Tylenchia	Tylenchida			+	+	
		Allantonematida				+	
		Myenchida				+	
		Aphelenchida			+	+	

orders are either exclusively parasitic or free-living, although a few (dorylaimids, rhabditids and diplogasterids) contain both parasitic and free-living species. The mermithids are parasitic as juveniles but have free-living adults (the reverse of the usual pattern). Parasitism appears to have arisen independently among several orders of nematodes (see Chapter 5).

2 FUNCTIONAL ORGANISATION

The Pseudocoel

The so-called pseudocoelomates are differentiated from the acoelomate groups by the presence of a true fluid-filled body cavity called a pseudocoel. Despite the name there is nothing false about this cavity, but it differs from a coelom in that it is bounded by the body-wall musculature and the gut wall; there are no muscles around the gut (Figure 2.1). Also the cavity is not lined with a membrane (no mesentery) and the organs, including the gut, lie free in the pseudocoel. There may, however, be a nucleated or fibrous network within the pseudocoel (Maggenti, 1976). The continuity of the pseudocoel has been demonstrated in *Ascaris* by the continuous perfusion of dye through the cavity (Fleming and Fetterer, 1984).

There is some doubt concerning the development of the pseudocoel. It was thought to be derived from the blastocoel of the embryo but is now thought to be a schizocoel formed by splits in the mesoderm during development (Maggenti, 1976). As discussed in Chapter 1, the pseudocoel may have different origins in the different pseudocoelomate groups and in different species of

Figure 2 1: The Basic Organisation of Acoelomates (A), pseudocoelomates (B) and coelomates (C). Redrawn from Crofton (1966)

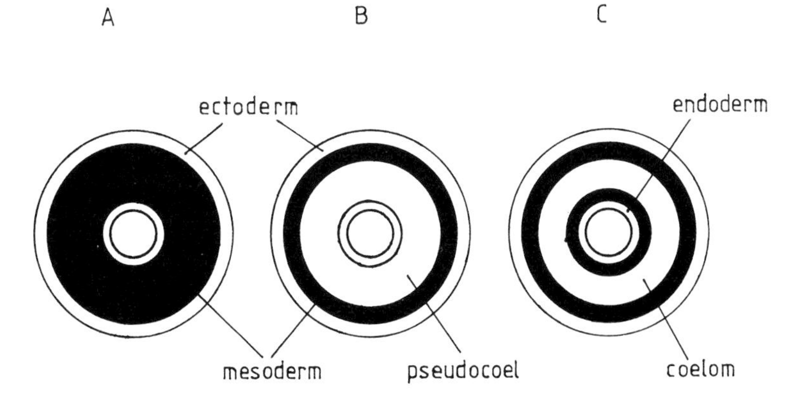

nematode, and thus does not provide a reliable character for including these groups in the same phylum.

The evolution of a body cavity was an essential development for the complexity and size of the body to increase above that of the acoelomate condition (Clark, 1967). In the absence of a circulatory system an animal is dependent upon diffusion through the tissues for the transport of nutrients, waste products and respiratory gases. The pseudocoel is in contact with all the tissues of the body, providing a primitive circulatory system. The long thin shape of most nematodes aids processes that depend upon diffusion across the body wall by decreasing the surface/volume ratio.

The presence of a body cavity also enables more complex locomotion. The tissues of an acoelomate cannot move relative to one another, limiting the types of muscular movement that can be made; acoelomates are thus restricted to ciliary creeping, looping and swimming (Clark, 1967). A body cavity isolates the body musculature from the other tissues and organ systems, allowing the muscles to be divided into segments and bands which allow more complex locomotion. The fluid-filled body cavity also provides a hydrostatic skeleton which plays a crucial role in the locomotion and functional organisation of nematodes (see below).

The pseudocoelomic fluid is a complex mixture. In *Ascaris* it contains lipid, protein, carbohydrates, nitrogenous compounds, inorganic ions and haemoglobin (von Brand, 1973). Haemoglobin may be found in the cuticle and intestinal wall of nematodes, as well as in the pseudocoel. It acts as an oxygen store and improves the efficiency of transport in conditions of low oxygen tension (Atkinson, 1976).

The Cuticle

General Organisation

With the exception of a few species parasitic in the haemocoel of arthropods, the entire surface of a nematode is covered by cuticle. The cuticle is multilayered and its study has suffered from confusion over the nomenclature of these layers. Bird (1980) has suggested a common nomenclature (Figure 2.2). The proximal parts of systems which open to the exterior (oesophagus, rectum, amphids, phasmids, reproductive and excretory systems) are also

Figure 2.2: The Structure of the Nematode Cuticle. Redrawn from Bird (1980)

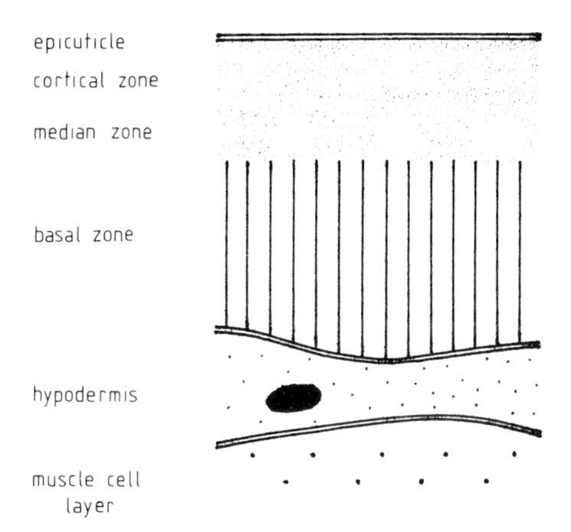

lined by cuticle, but this has a different structure to the surface cuticle.

The epicuticle is usually a triple-layered structure on the outer surface of the cuticle. Beneath this are the cortical and median zones. These may be traversed by various struts, fibres or circular lamellae, which are thought to act as skeletal supports (Bird, 1971). The median zone of the cuticle of adult *Nippostrongylus brasiliensis* is fluid-filled and is the site of the cuticular haemoglobin found in this species (Lee, 1969). In other species the median zone appears homogeneous and may simply have a structural function. The basal zone forms the inner layer of the cuticle and is fibrous or consists of material which, when sectioned in some orientations, appears striated. Beneath the cuticle, and connecting it to the muscle cells, is a cellular or syncytial hypodermis.

The structure of the cuticle varies between different species, and between different stages of the same species, reflecting its function in different environments. Its structure, however, reflects the basic pattern, which can be modified by the omission or addition of layers or by their subdivision.

Surface features in various parts of the body are formed by elaborations of the cuticle (Bird, 1971). The most functionally important of these are the annulations and the lateral alae. Annulations are transverse markings on the dorsal and ventral surfaces (Plate 2.1), which are absent at the lateral alae in most species and give a false appearance of segmentation. Annulations are formed

Plate 2.1: Scanning Electron Micrograph of the Infective Juvenile of *Trichostrongylus colubriformis*. The annulations (a) and lateral alae (la) of the cuticle (sheath) can be seen. × 1500. Reproduced from Wharton (1982a) with permission

Plate 2.2: Tangential Section through the Striated Basal Zone of the Cuticle of *T. colubriformis* Infective Juvenile. The striated layer appears to consist of two sets of parallel laminae which intersect at right angles. Electron-dense spots can be observed at alternate intersections (arrow). × 135 000. Original

Plate 2.3: The Epicuticle (e) of *T. colubriformis* Infective Juvenile. The epicuticle is complex in this species, consisting of nine distinct layers. In most species it is a three-layered structure. × 150 000. Original

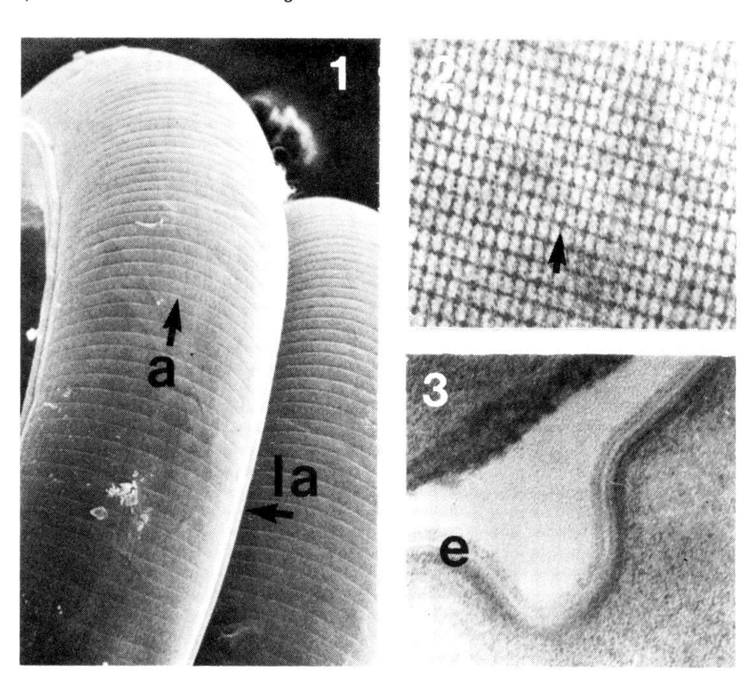

by a decrease in the thickness of the cortical and/or median zones of the cuticle. They enable the nematode to change in shape and the cuticle to flex during dorsoventral contractions.

The lateral alae overlie the lateral hypodermal cords and run longitudinally along the whole length of the body. They have a complex structure quite different from that of the general body cuticle, and they may provide a degree of longitudinal stiffening and allow for changes in the worm's diameter. Since nematodes lie on their sides, the lateral alae are in contact with the substrate and, therefore, increase the efficiency of locomotion by increasing

Plate 2.4: Longitudinal Section through the Excretory Ampulla (ea) and Excretory Cell (ec) of the Infective Juvenile of *Haemonchus contortus*. The excretory ampulla is surrounded by a complex of membranes, which have been called canaliculi. The excretory cells are packed with electron-dense granules. × 15 000. Reproduced from Wharton and Sommerville (1984) with permission

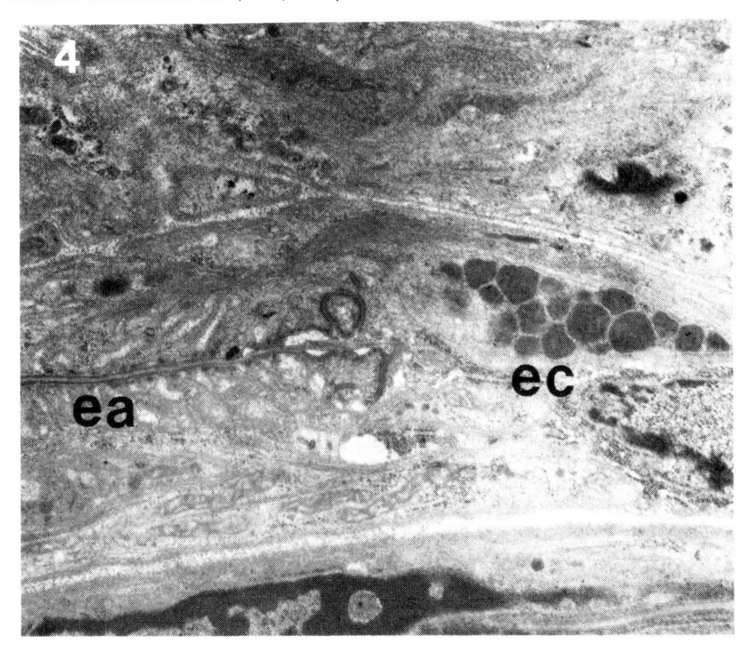

friction and preventing rolling. In forms with lateral alae which are very extended, they may act as fins during swimming (Lee, 1969).

The Basal Zone

The structure of the basal zone is often complex and plays an important role in nematode biology. In ascarids and some other nematodes it consists of three layers of parallel fibres, which act in conjunction with the turgor pressure of the pseudocoel to provide an antagonistic system to the longitudinal muscles (see below). The basal zone has a striated appearance in the infective juveniles of both animal and plant parasitic nematodes, some free-living nematodes and in adult trichurids. The periodicities observed are different in transverse and longitudinal sections. In tangential sections the layer appears as a lattice of intersecting lines with electron-dense spots at alternate intersections (Plate 2.2).

There have been a number of suggestions as to how this layer is constructed. Wisse and Daems (1968) suggest rods interconnected by membranes, whereas Popham and Webster (1978) consider it as a series of intersecting laminae. I have recently suggested (Wharton, 1982b) a development of the Popham and Webster (1978) model which shows how this layer might be constructed from a single repeating unit (Figure 2.3) and may be a crystalline structure. Crystalline proteins are responsible for a striated appearance in a number of biological structures and inclusions. Inglis (1983b) has similarly suggested that the striated basal zone is analogous to a liquid crystal and that this, and other cuticular structures, forms from an amorphous precursor by a process of self-assembly.

A striated basal zone is found mainly in nematodes which are exposed to hazardous environments and it may be responsible for the resistance of these stages to chemicals and desiccation.

The Epicuticle

The epicuticle is usually a triple-layered structure but can consist of as many as nine distinct layers (Plate 2.3). The unit membrane-like appearance of this layer and its lipoprotein composition have led to the suggestion that it has much in common with a cell mem-

Figure 2.3: The Striated Layer Can Be Constructed from a Single Unit Consisting of a Block with Opposite Corners Missing (Bottom). This unit can be arranged with an electron-dense matrix between the blocks to give the pattern observed in tangential sections (top) and the periodicities observed in transverse (a), longitudinal (b) and oblique (c) sections (original)

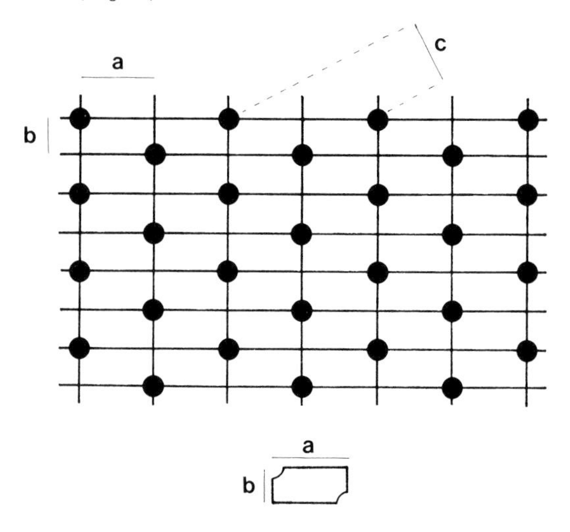

brane. The epicuticle lining the rectum of *Meloidogyne javanica* is continuous with the cell membrane lining the rectal gland (Bird, 1980). Carbohydrate residues are present on the surface of a number of animal cells and their presence on the surface of nematodes has been demonstrated using oligosaccharide-specific lectins. These include galactose, glucose, mannose, N-acetylglucosamine (Zuckerman, Kahane and Himmelhoch, 1979), sialic acid and perhaps N-acetylgalactosamine (Spiegel, Cohn and Spiegel, 1982). The inability of proteases to reduce labelling indicates that these residues occur as glycolipids rather than glycoproteins or that the carbohydrate residues are protected from attack.

There are indications that a surface coat or glycocalyx overlies the epicuticle (Bird, 1980), and labelling with cationised ferritin indicates a negative surface charge. Recent work with ultracryosections, however, indicates that negatively charged groups are distributed throughout the cuticle (Himmelhoch and Zuckerman, 1983).

Adenosine triphosphatase, an enzyme commonly associated with mammalian cell membranes, is absent from the epicuticle of *Trichinella spiralis* infective juveniles (Jungery, Mackenzie and Ogilvie, 1983). Freeze-fracture images of nematode epicuticle lack the membrane particles common in those of cell membranes. Thus, although the epicuticle has some of the characteristics of a cell membrane, there are also many differences. Locke (1982) considers that it is incorrect to refer to the epicuticle as unit membrane-like, and suggests that it, and similar structures in other organisms, should be called an envelope.

Perhaps of more importance than the question of whether the epicuticle is or is not a unit membrane is the realisation that it is a dynamic structure. Surface labelling of *T. spiralis* with ^{125}I indicates changes in epicuticular components between different stages (Philipp, Parkhouse and Ogilvie, 1980). There are also quantitative changes during the maturation of adult worms. Infective juveniles and adults shed the labelled component when incubated in normal rat serum and other media. There are also several reports of the shedding of antibody complexes formed to epicuticular antigens (Murrell and Graham, 1983). If nematodes can continually replace the epicuticle, this may provide a mechanism for the evasion of the immune response of a host to surface components. It may also indicate how the cuticular permeability barrier is maintained and repaired.

Chemical Composition

The cuticle consists mainly of protein and lipid with associated mucopolysaccharides. Collagen is the dominant structural protein. This has rather different properties to vertebrate collagens and in most cases lacks its characteristic 64 nm banding pattern (Barrett, 1981). It belongs to the group called secreted collagens (Woodhead-Galloway, 1980). In *Ascaris* the individual polypeptide chains are folded back on themselves to form hydrogen-bonded, three-stranded superhelices, which are held together by sulphydryl bridges to form the collagen subunit (Figure 2.4). No other collagen has been found with this structure or is stabilised by sulphydryl bridges. A second structural protein, cuticulin, containing dityrosine linkages has been found in *Ascaris* cuticle.

Little is known about the lipid component of the cuticle. Lipid, in particular that in the epicuticle, must play an important part in the control of cuticular permeability.

Functions of the Cuticle

The cuticle plays a number of important roles in the functional organisation of nematodes. In conjunction with the high turgor pressure of the pseudocoel it provides a hydrostatic skeleton and the antagonistic system to the longitudinal muscles. It also provides a degree of mechanical protection. Collagen fibres are inextensible and have a high resistance to longitudinal strain. The fibres are randomly orientated, or in parallel layers of different orientations, and thus provide resistance to stress and strain in different directions. The constituents of the cuticle are chemically inert and the cuticle itself is resistant to chemical attack. It has been suggested that the cuticle is quinone-tanned but there is little direct evidence for this, apart from in the cyst wall of *Globodera* (Awan and Hominick, 1982).

The selective permeability of the cuticle controls the exchange of materials between the nematode and the environment through

Figure 2.4: Proposed Structure of *Ascaris lumbricoides* Cuticle Collagen. Each polypeptide chain is back-folded to form a three-stranded hydrogen-bonded superhelix. Thick bars represent sulphydryl bridges and thin lines the triple helix region. Redrawn from Barrett (1981).

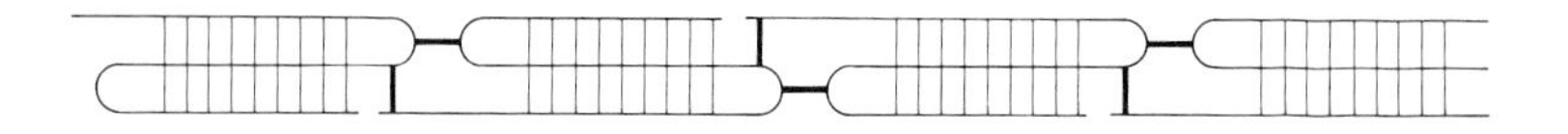

the body wall. The preparation of nematodes for electron microscopy is often hampered by the extremely limited permeability of the cuticle, and some species are still swimming around after several hours in fixative. There have been few studies on cuticular permeability and its control despite the importance of these properties in the penetration of nematicides and anthelmintics.

Plant parasitic and fungivorous nematodes are more easily penetrated by non-polar, lipid-soluble substances (Marks, Thomason and Castro, 1968; Castro and Thomason, 1973). The selective permeability is destroyed at high temperatures. These observations suggest that the permeability barrier is a lipid layer, probably the epicuticle. Transcuticular uptake of nutrients occurs in some insect parasites and adult filarids, but these nematodes are surrounded by the blood or lymph of their hosts and are perhaps atypical. Some uptake of glucose and alanine through the cuticle of ascarids has been demonstrated but the rates are too low to be physiologically significant (Barrett, 1981).

Evidence is accumulating that the cuticle is not just an inert structure but is physiologically active and is perhaps repaired and maintained by the underlying hypodermis. The permeability of the cuticle of anabiotic nematodes (which can survive complete desiccation and the cessation of metabolism — see Chapter 7) is greater when dry but the permeability barrier is re-established after rehydration. Differences in the structure of the epicuticle reflect the resistance of the nematode to environmental hazards (Bird, 1980). Many of the life processes of nematodes may be controlled by changes in the properties of the cuticle.

The Turgor Pressure System

In order to produce movement, muscular contraction needs something to act against. Nematodes lack a rigid skeleton and most species collapse due to water loss when exposed to air. The pseudocoel, in conjunction with the limitations on expansion imposed by the cuticle, acts as a hydrostatic skeleton which maintains body shape and provides the reaction to the contraction of the longitudinal muscles (Harris and Crofton, 1957). Many invertebrates possess hydrostatic skeletons (Barrington, 1979). The nematode system is unique because of the high turgor pressures recorded and because it lacks circular body wall muscles.

The Hydrostatic Skeleton

The presence of a high turgor pressure has been demonstrated by direct measurement in *Ascaris.* Harris and Crofton (1957) used a pressure gauge connected to the pseudocoel via a cannula. They recorded pressures of 16-225 mmHg with a mean of 70 mmHg. This is much higher than the internal pressures recorded in most other invertebrates (Table 2.1). A method for the measurement of turgor pressure in small nematodes has not yet been developed. However, most species appear to possess a high internal pressure because if the body is cut the body contents are rapidly ejected through the wound.

The hydrostatic skeleton gives rigidity and, acting like the fluid in an hydraulic brake, transmits pressure increases due to the contraction of the longitudinal muscles to other parts of the body. Being incompressible, an increase in pressure will be directly transmitted throughout the system. Contraction in one part of the system must, therefore, result in expansion elsewhere. To provide an antagonistic system to the longitudinal muscles, these forces must be resolved longitudinally (Figure 2.5). This is ensured by the structure and mechanical properties of the cuticle.

Cuticular Fibre Systems

The basal zone of the adult cuticle of *Ascaris*, and of many other species, consists of three layers of parallel fibres. The fibres are parallel within each layer but the fibre direction of the middle layer is at an angle to that of the inner and outer layers. The fibres therefore form two helices which spiral round the body in opposite directions (Figure 2.6). The sheaths of the infective juveniles of

Table 2.1: Internal Pressures Recorded in Various Invertebrates

Organism	Pressure (mmHg)
Ascaris lumbricoides	16-225
Potamobius fluviatilis	10.5-18
Carcinus maenas	3.5-19
Peripatopsis sp.	2-15
Arenicola sp.	10-20
Calliactis parasitica	0-10
Lumbricus terrestris	1.5-21
Balanus nubilus	200-430

From Crofton (1966)

Figure 2.5: The Possible Consequences of the Contraction of the Longitudinal Muscles of Nematodes. When the longitudinal muscles A-B (1) contract, the section B-C may respond by becoming wider (2) or by lengthening (3). For the turgor pressure system to act antagonistically to the longitudinal muscles, the properties of the cuticle must be such that (3) occurs. Original

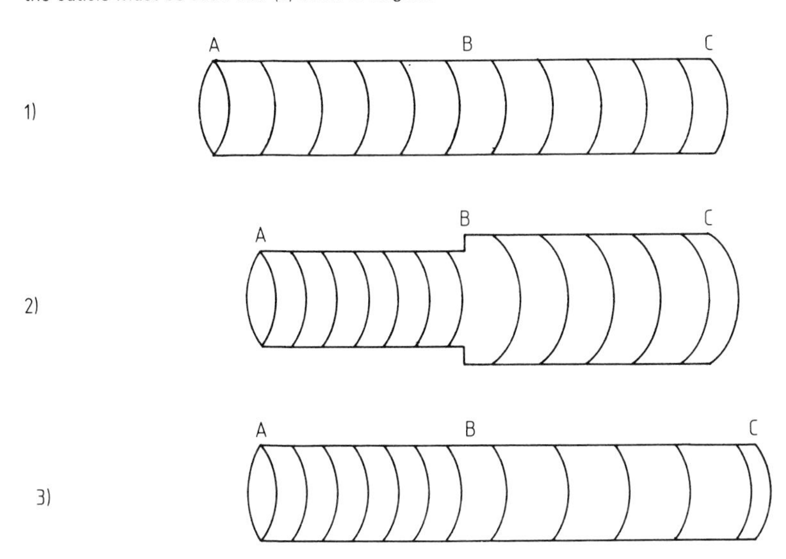

trichostrongyle nematodes also possess this arrangement (Figure 2.7).

This structure has been described as a 'trellis-work' or 'lazy-tongs' arrangement (Harris and Crofton, 1957; Crofton, 1966). However, this is not a good analogy as there is no interweaving or interconnections between the fibres. The fibres are organised into layers and can only interact with fibres of a different orientation at the boundaries of these layers.

A more easily understood model is that of Cowey (1952) for a similar system found in nemertines and most recently applied to nematodes by O'Grady (1983). Cowey considered the properties of the cylinder enclosed by a single turn of a spiral fibre, of length FB (Figure 2.8). As the length of the body changes, the angle the fibre makes with the longitudinal axis of the body (the spiral angle, θ) changes. The body can get longer and thinner or shorter and thicker. The length (l), radius (r) and the volume ($v = \pi r^2 l$, assuming a cylinder) are dependent upon the spiral angle. As θ tends to 0, then l tends to FB, r tends to 0, and v tends to 0. As θ tends to 90°, then l tends to 0, r tends to FB/2 and v tends to 0.

Figure 2.6: The Fibres of the Basal Zone of the Cuticle Form Two Helices which Spiral Round the Body in Opposite Directions. Redrawn from Cowey (1952)

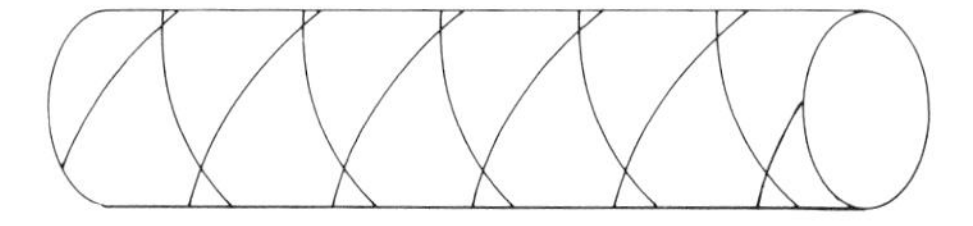

Figure 2.7: The Structure of the Sheath of the Infective Juvenile of *Trichostrongylus colubriformis*. The basal zone consists of three layers of spirally wound fibres. EP, epicuticle; CZ, cortical zone; BZ, basal zone. Original

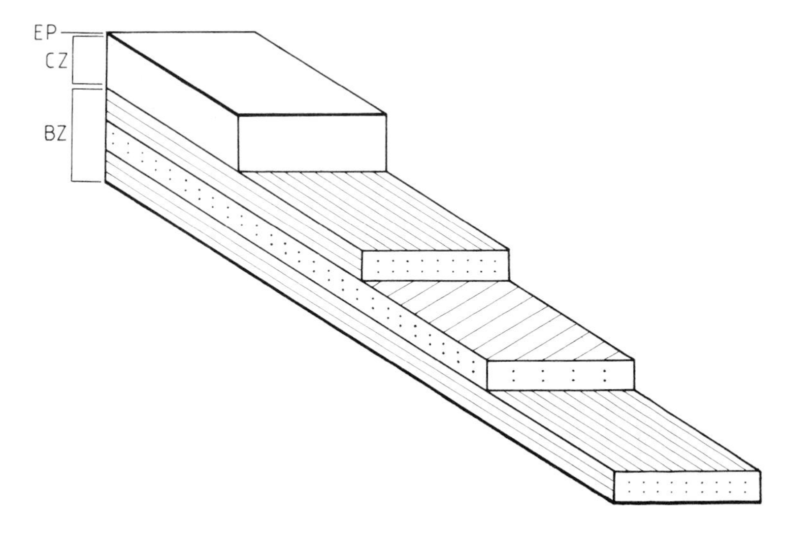

Figure 2.8: A Spiral Fibre of Length FB and at a Spiral Angle θ to the Longitudinal Axis of the Body Encloses a Cylinder of length *l* and radius *r*. The volume of the cylinder ($\pi r^2 l$) is dependent on the spiral angle θ. Redrawn from Cowey (1952)

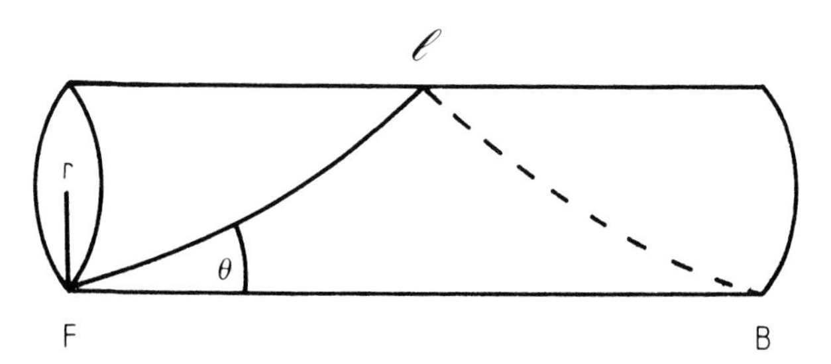

Figure 2.9: The Cylinder of Figure 2.8 Cut and Laid Flat (A). As θ tends to 0, then *l* tends to FB, 2π*r* tends to 0, and the volume tends to 0 (B). As θ tends to 90°, then *l* tends to 0, 2π*r* tends to FB and the volume tends to 0 (C). Original

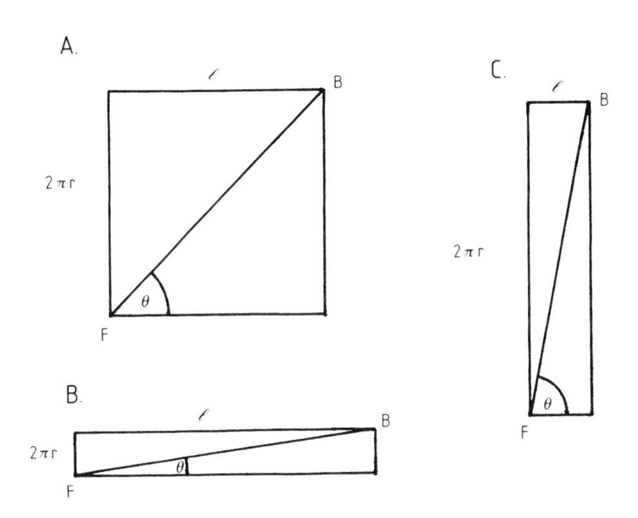

This can be more clearly seen if the cylinder is cut along the top and laid flat (Figure 2.9).

The change in volume with a change in spiral angle can be calculated (Figure 2.10) and maximum volume occurs at a spiral angle of $\theta = 54°44'$. At angles less than this, decreasing the length increases the volume. At angles greater than this, decreasing the length decreases the volume.

Assuming a closed system, pressure varies in inverse proportion to volume (Figure 2.10). At $\theta > 54°44'$, therefore, a decrease in length results in an increase in pressure. The spiral angle in *Ascaris* has been measured to be 75°30′ (Harris and Crofton, 1957). A decrease in spiral angle below the critical value of 54°44′ due to body lengthening is not likely to occur because of the restriction on volume reduction imposed by the body contents and the cuticle.

The fibre system thus ensures that contraction of the longitudinal muscles causes an increase in turgor pressure and that this results in a lengthening of the body elsewhere. The turgor pressure system thus provides the antagonistic action to the longitudinal muscles in the absence of circular muscles.

Figure 2.10: Graph Showing How Volume (*V*) and Pressure (*P*) vary with spiral angle (θ) and length (*l*). Maximum volume and minimum pressure occur at the spiral angle of 54°44′. Redrawn from O'Grady (1983)

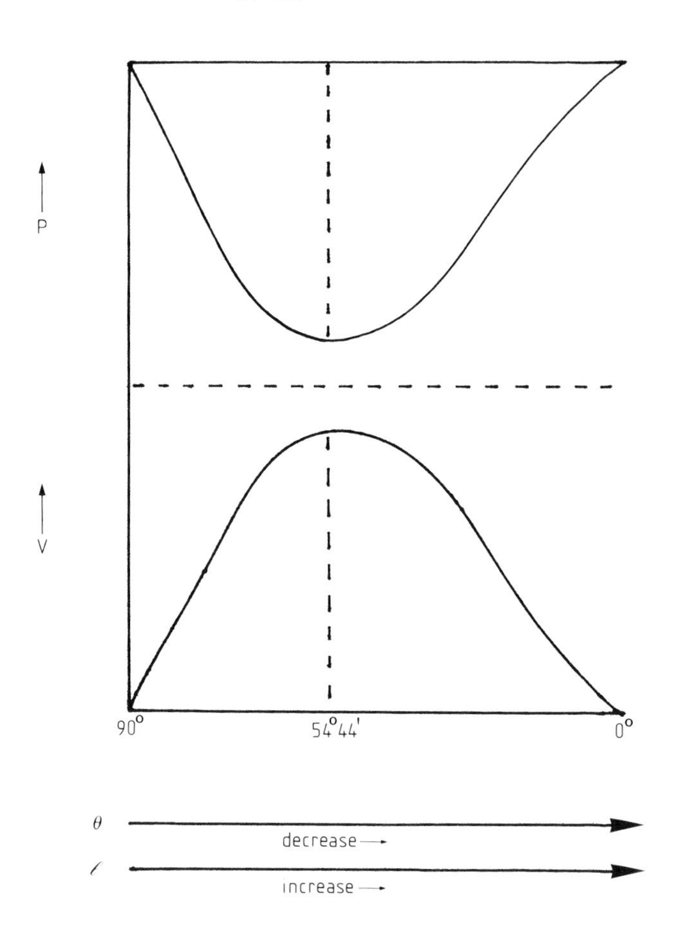

Generality of the Turgor Pressure System

Many small nematodes and juvenile stages appear to lack spirally wound fibre systems. This has led to suggestions that the turgor pressure system does not apply to all nematodes (e.g. O'Grady, 1983). The system as described by Cowey (1952) will, however, function with only a single layer of spirally wound fibres as well as with the three layers that form a double spiral in *Ascaris*. A single layer of fibres would be difficult to demonstrate in a small nematode, unless special techniques such as freeze-etching were used. Indeed, unless there is a marked contrast between the fibres and their surrounding matrix, they would not be resolved as fibres at all in thin sections and the layer would appear amorphous. Spirally wound fibre systems may therefore be more widely present than is realised.

There have been suggestions as to how the cuticle might participate in a turgor pressure system in the absence of spirally wound fibres. All that is required is that the physical properties of the cuticle are such that lengthening, rather than an increase in diameter, is favoured. Cuticular annulations and other surface markings may act in this fashion (Inglis, 1964), and the striated basal zone found in the cuticle of many juvenile stages could provide a progressive resistance to deformation (Wisse and Daems, 1968).

In electron micrographs of many small nematodes the pseudocoel appears to be absent or very much reduced and the body appears to be completely filled by the muscles and intestine. This may be a fixation artefact, due to the swelling of tissues, as the ability of organ systems to move relative to one another in many species suggests that some sort of cavity is present. Seymour (1973) has suggested that, if a cavity is absent, the body contents themselves provide the deformable material of the hydrostatic skeleton. A few species can continue to move even if the body wall is punctured and, therefore, may not rely on a turgor pressure system for locomotion or contain partitions within the pseudocoel that limit the effects of such damage.

Despite these difficulties it appears that most nematodes rely on a high turgor pressure for locomotion. This has important consequences for other aspects of their functional organisation.

General Body Form

The high internal turgor pressure may be responsible for the remarkable constancy of nematode body form (Crofton, 1966). If the cuticle were freely extensible in all directions, the body would take the shape with the minimum surface area, a sphere. However, the cuticle limits lateral expansion and the body takes the form of a cylinder. A cylindrical body form is also suitable for the sinusoidal locomotion of nematodes, and for an animal that relies mainly upon diffusion for the transport of respiratory gases, because it has a relatively high surface/volume ratio (Atkinson, 1980).

Very few nematodes depart from this cylindrical body form. Those that do are nearly all spherical or oval. An example is the mature female of the potato cyst nematode, *Globodera rostochiensis.* The eggs are stored within the body of the spherical female, which dies when gravid. The cuticle then tans and hardens to form a protective cyst (Awan and Hominick, 1982).

The restrictions imposed by the turgor pressure system have limited the extent to which nematodes have altered their body form to adapt to different environments, as is seen in other groups. Dependence on a turgor pressure system also limits their activity to environments where they are enclosed by at least a film of water; the turgor pressure is lost if the nematode loses water through desiccation. Despite these limitations nematodes are a very successful group and have exploited a great variety of habitats.

The Alimentary Canal

In most species the alimentary canal is a straight tube which opens anteriorly at the mouth and posteriorly at the anus. It consists of three main regions: the stomodaeum (mouth, buccal capsule and oesophagus), the intestine and the proctodaeum (rectum or cloaca and anus). Both the proctodaeum and the stomodaeum are lined by cuticle, which is shed at the moult. The intestine consists of a cylinder, composed of a single layer of cells with microvilli on their luminal surface. Some species possess outpushings, called diverticulae or caecae, in the anterior region of the intestine. These may increase the area available for enzyme secretion and food absorption.

The functioning of the alimentary canal is intimately linked to the turgor pressure system. The intestine collapses under the turgor pressure and can only be filled by a pump, the muscular oesophagus, which overcomes this pressure. Filling the intestine increases the turgor pressure which provides the force for defecation. Defecation is controlled by the opening of the anus.

The intestine does not possess any muscles, although it may have contractile elements, and it is not capable of peristalsis. Movement of food along the gut results from the pumping of the oesophagus and from the activity of the body muscles, which is transmitted to the gut through the pseudocoel. The only muscles associated with the alimentary canal are those of the oesophagus and the anus (anal dilator muscles).

Feeding and the Oesophageal Pump

Nematodes are mainly liquid feeders. They feed on liquids directly (blood, plant juices) or on particles ingested in a liquid medium (bacteria, tissue fragments). The oesophageal pumping mechanism is well adapted for ingesting such material.

The structures associated with the anterior part of the stomodaeum and the mouth are complex and variable, reflecting the different feeding habits of the species concerned. Wright (1976) has provided useful definitions of the structures involved. Lips are movable projections surrounding the mouth which assist in feeding, and the buccal capsule is the part of the anterior stomodaeum which is modified for the ingestion of food. There is a marked difference between the ultrastructure of the body cuticle and of that lining the oesophagus. This allows us to distinguish between the stoma (the region posterior to the oral opening which is lined by body cuticle) and the rest of the buccal capsule (which is lined by oesophageal cuticle).

Some species do not have any body cuticle associated with the buccal capsule and, therefore, lack a stoma (they are astomatous). The cuticle lining the buccal capsule may be modified as plates or teeth which are used by animal parasitic nematodes for attachment, or to abrade tissue prior to ingestion. Plant parasitic nematodes, the juveniles of mermithids and some predacious and mycophagous species possess a stylet (Figure 2.11). This is a hypodermic-like structure derived from the walls of the oesophagus or buccal capsule (Bird, 1971). The stylet can be extended and retracted by its attached musculature and is used by plant parasites both to inject enzymes into a plant cell and to ingest the partly-digested contents. Oesophageal pumping is associated with both of these processes.

The structure of the oesophagus is very variable and it may consist of a variety of bulbs and isthmuses (Figure 2.12). Some species possess valves that prevent the oesophagus emptying to the outside or filling from the intestine. The opening and closing of these valves is co-ordinated with the pumping cycle. In the absence of valves a wave of dilation passes down the oesophagus so that only

Figure 2.11: Diagram of the Stylet of *Aphelenchoides blastophthorus*. clp, cuticular lining of procorpus; co, cuticular collar; gn, gustatory neuron; pr, proprioceptor; prc, procorpus; sc, stomatal cuticle; sco, stylet cone; sk, stylet knob; sl, stylet lumen; smu, stylet protractor muscle; so, stylet orifice; ss, stylet shaft; stc, stomatal cavity; tc, electron-translucent cuticular layer; 5, layer 5 of the stylet

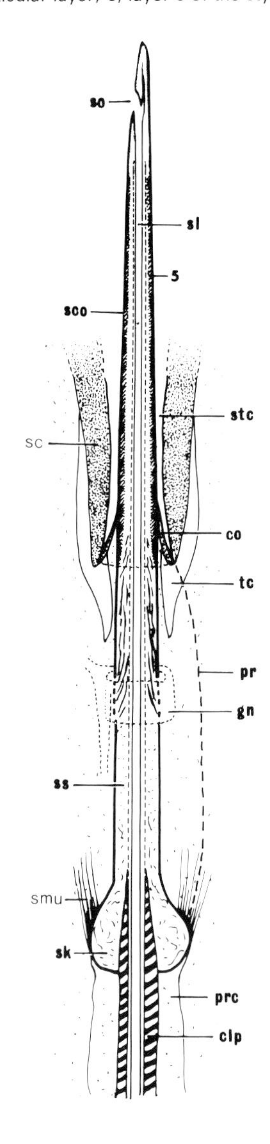

Reproduced from Shepherd, Clark and Hooper (1980) with permission

Figure 2.12: Variation in the Structure of the Nematode Oesophagus. A, *Enoplus*; B, *Agamermis*; C, *Leidynema*; D, *Ascaris*; E, *Oxyuris*; F, *Aplectana*. Redrawn from Chitwood and Chitwood (1974); Mapes (1965)

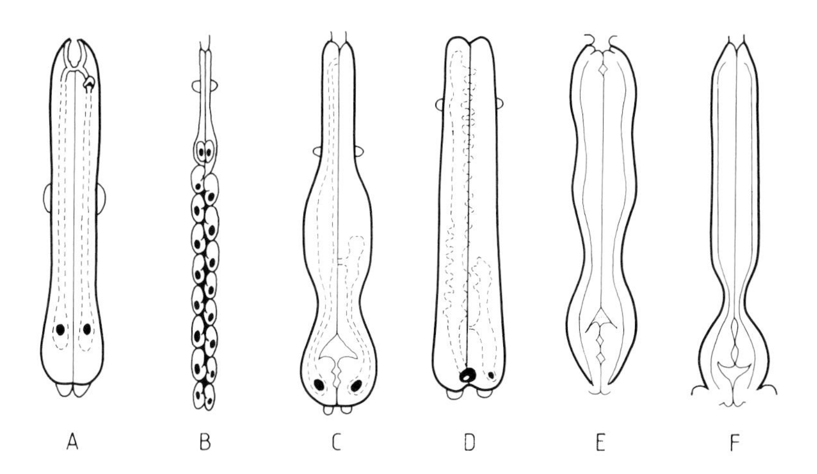

half its length is dilated at any one time (Mapes, 1965); this provides a self-sealing system. In addition to pumping, the oesophagus may be involved in filtering and food maceration.

Bennet-Clark's (1976) attempt to explain the pumping mechanism of the oesophagus overcomes some of the problems associated with earlier models. The oesophagus has a triradiate cuticle-lined lumen which is connected to the outer wall by radial muscles. The oesophageal cuticle has the properties of an elastic protein similar to resilin. The oesophagus can be considered as two concentric cylinders (the outer wall and the lining of the lumen) which are connected by the radial muscles. Such a cylinder increases in diameter more easily than in length. When the radial muscles contract, there is an increase in pressure within the cylinder. If this pressure is less than that within the pseudocoel, the excess pressure prevents the cylinder increasing in diameter and it therefore lengthens and becomes thinner, keeping the lumen closed (Figure 2.13). As the radial muscles continue to contract, the pressure within the cylinder increases until it is greater than that in the pseudocoel. The cylinder then increases in diameter and minimises stresses by becoming shorter and wider, resulting in the opening of the lumen (Figure 2.13).

When the radial muscles relax, or if the pressure within the cylinder is decreased by the opening of the oesophageal-intestinal

Figure 2.13: The Oesophageal Pumping Mechanism Proposed by Bennet-Clark (1976). (A) Muscles relaxed. (B) Muscles contracted. The pressure within the pseudocoel exceeds that within the oesophagus; it lengthens and does not dilate, keeping the lumen closed. (C) The pressure within the oesophagus exceeds that within the pseudocoel. It shortens and widens, opening the lumen, L_0 resting length, A_0 resting area of cross-section, D_0 resting diameter, V_0 resting volume of musculature

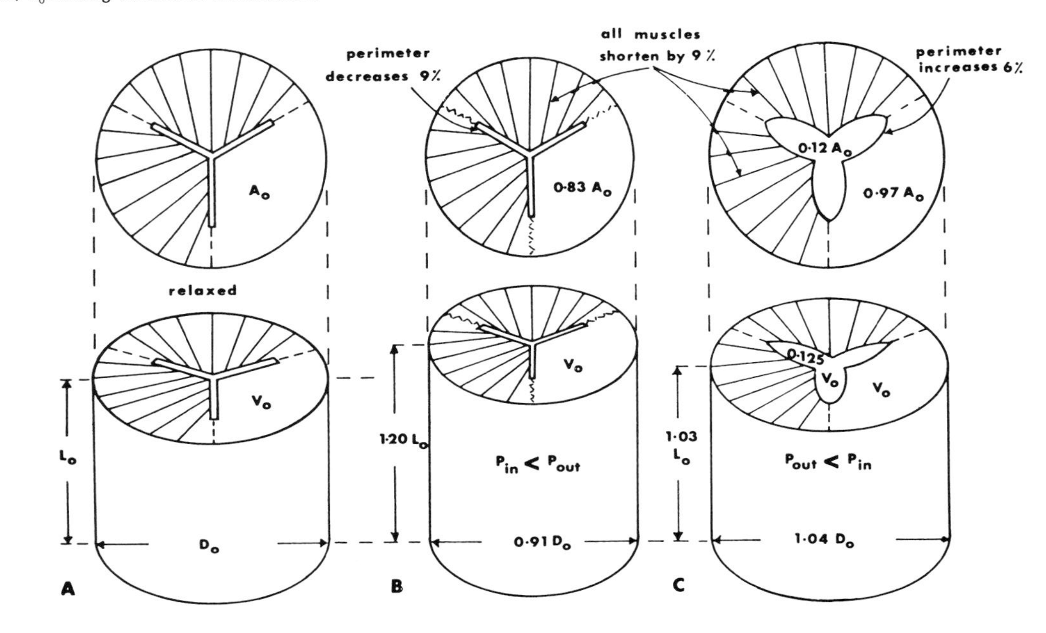

Reproduced from Bennet-Clark (1976) with permission

valve, the process is reversed. This, together with the elasticity of the cuticular lining, closes the lumen. Bennet-Clark (1976) has also shown that this system works best with three-rayed rather than two- or four-rayed pumps. All nematodes so far examined possess a triradiate oesophagus.

The Bennet-Clark model has been criticised by Roggen (1982), who prefers his earlier hoop-strain model. This model assumes that the outer wall of the oesophagus can be considered as an isotropic elastic membrane loaded by internal pressure: the hoop stress is therefore twice the longitudinal stress. The model also assumes that the Poisson ratio for this material is 0.5, and under these conditions the longitudinal strain is zero. The outer wall of the oesophagus can therefore only respond to pressure changes within the cylinder by changing in diameter. As the inner wall of the oesophagus is attached to the outer wall by the radial muscles, this will result in the opening or closing of the lumen.

Pulsation rates as high as 4 per second have been recorded in *Ascaris* (Mapes, 1965). Rhythmic action potentials associated with the contractions of the oesophagus have been demonstrated (del Castillo, de Mello and Murales, 1964), and the oesophageal nervous system is more or less autonomous (Albertson and Thomson, 1976). The oesophagus may act as a single, electrically excitable unit whose contractions are under myogenic control.

Seymour (1983) suggests that the oesophageal pump of *Ditylenchus dipsaci* acts as a myogenically controlled click mechanism. The pump lining may have two stable positions, open and closed, and is 'clicked' between them by the action of the radial muscles. Click mechanisms were first described in the wing bases of insects. They allow much greater frequencies of movement than could be achieved by direct motoneuron control.

Digestion and Absorption

The intestine consists of a central lumen, surrounded by a single layer of cells, which may form a syncytium. In non-feeding juveniles and adults (e.g. *Trichostrongylus colubriformis* J3, adult mermithids), the lumen may be eliminated and the intestinal cells modified for food storage. Food is stored as glycogen, lipid and protein (Bird, 1971). In most species, however, the intestine is involved in digestion and absorption.

The luminal surfaces of the intestinal cells are covered by finger-like microvilli (Figure 2.14). These increase the surface area

Figure 2.14: The Syncytial Intestine of *Haemonchus contortus*

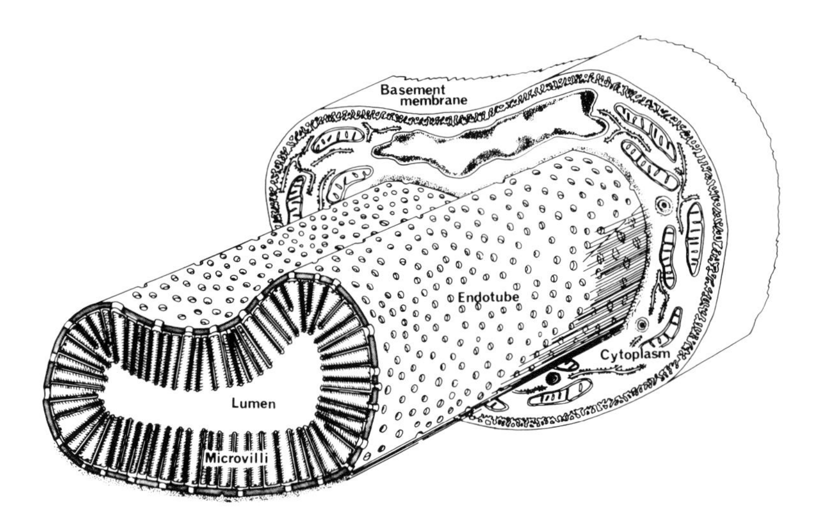

Reproduced from Munn and Greenwood (1983) with permission

available for absorption, in a similar manner to the microvilli of the mammalian intestine. The microvilli are covered by a glycocalyx which is implicated in the secretory activities of the intestine and may be the site upon which the digestive enzymes are absorbed (Lee and Atkinson, 1976).

A helical protein called contortin is associated with the glycocalyx in *Haemonchus contortus* and *Ostertagia circumcincta* (Munn, 1977). This is thought to act as an anticoagulant in these blood-feeders. *Bradynema* sp., a parasite in the haemocoel of insects, lacks a functional gut. It also has no cuticle, possesses microvilli on the body surface and probably takes up nutrients through this surface (Riding, 1970).

Little is known about how absorption occurs in the nematode intestine. There is evidence of both active and passive transport across the gut of *Ascaris* (Lee and Atkinson, 1976). A thickened layer beneath the microvilli in syncytial intestines enables an almost pure brush-border preparation to be dissected (Munn and Greenwood, 1983). Such preparations will hopefully tell us much about the functioning of the microvilli and the glycocalyx.

A number of digestive enzymes, including proteases, lipases, and carbohydrases, have been found in the intestine (Barrett, 1981). These are associated with the microvilli and may be secreted by the intestinal cells. Enzymes are also secreted by the oesophageal glands and are involved in extracorporeal digestion.

A number of free-living nematodes can be kept in axenic culture using chemically defined media (Vanfleteren, 1980). Their nutritional requirements can be investigated by omitting ingredients from these media or by adding anti-metabolites. These nematodes have essential requirements for a number of vitamins, amino acids, lipids and carbohydrates.

Defecation

A high internal turgor pressure provides the force for defecation. Indeed, liquid faeces may be ejected for some considerable distance upon the opening of the anus. An intestinal-rectal valve controls the entry of material into the rectum. In *Aphelenchoides blastophthorus* the intestinal-rectal valve consists of convoluted, closely apposed membranes which are perhaps held together by intermolecular forces (Seymour and Shepherd, 1974). When the anus is opened by the contraction of the anal dilator muscles, the rectum empties under the force of the turgor pressure. When the muscles relax, the turgor pressure causes the anus to close. In males the rectum and the reproductive system have a joint opening in the cloaca.

A number of species possess rectal glands which empty their secretions via the rectum. In *Meloidogyne javanica* these are responsible for the secretion of the gelatinous matrix which encloses the egg mass (Bird, 1971).

The Excretory System

The nematode excretory system has been so named on morphological evidence only. Its primary role appears to be secretion and osmotic and ionic regulation. Urea is produced by a few species but ammonia is the major nitrogenous waste product (Wright and Newall, 1976). This is probably excreted via the gut rather than the excretory system.

Although physiological evidence is lacking, a number of structures have been suggested as the site of excretion. These include the gut, the cuticle, the bacillary band cells of trichurids (hypodermal gland cells on the anterior part of the body which open to the exterior via a cuticular pore) and the excretory system. Pseudocoelomocytes are ovoid or branched cells which occupy a fixed position within the pseudocoel. Their function is unclear but they may be involved in excretion and are known to accumulate vitamin B_{12} (Barrett, 1981). It has also been suggested that nitrogenous waste is eliminated along with the cuticle at a moult.

The excretory system may be either glandular or tubular (Bird, 1971). The glandular system consists of a single large cell, the ventral gland or renette, which opens to the exterior via a duct. The tubular type consists of two lateral ducts, lying next to the lateral cords and connected via transverse ducts to a cuticle-lined median duct or ampulla, which opens to the exterior via an excretory pore (Figure 2.15). One or two subventral glands or excretory cells may be associated with the median duct.

There have been a number of reports of material being released through the excretory pore, and secretory granules have been demonstrated in the ventral glands of a number of species. Enzymes may be secreted via the excretory system and involved in exodigestion and the histolytic penetration of host tissues. Some

Figure 2.15: The Excretory System of the Infective Juvenile of *Haemonchus contortus*. EA, excretory ampulla (median duct); EC, excretory cell; EP, excretory pore; EV, excretory valve; LD, lateral duct; TD, transverse duct

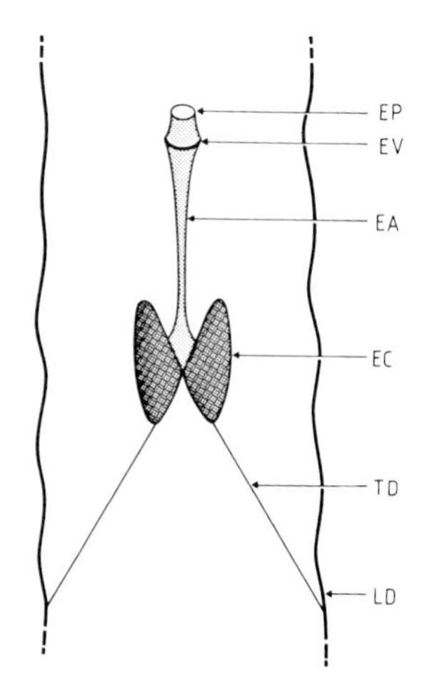

Reproduced from Wharton and Sommerville (1984) with permission

animal parasitic nematodes secrete acetylcholinesterase which may act as a 'biochemical holdfast' by damping down host gut peristalsis (Wright and Newall, 1976). The excretory system may also be the source of moulting enzyme and is implicated in the exsheathing process of strongyle infective juveniles (Wharton and Sommerville, 1984).

The tubular type of excretory system may be involved in osmotic and ionic regulation. The lateral and median excretory ducts are surrounded by fine tubules or canaliculi (Plate 2.4). These systems are similar in structure to the contractile vacuoles of protozoa and may similarly produce a hyposmotic fluid by the active secretion of ions (Wharton and Sommerville, 1984). The excretory system of nematodes is not like the protonephridial systems of other invertebrates as motile cilia (kinocilia) are absent. Harris and Crofton's (1957) suggestion that it may act as a high-pressure filtration system is also not borne out by structural evidence (Wharton and Sommerville, 1984).

The release of material via the excretory pore may be controlled by an excretory valve. This functions in a similar way to the anus. It is opened by the contraction of its dilator muscles and closed by the action of the turgor pressure (see Chapter 7).

The Reproductive System

Nematodes have a simple tubular reproductive system, which is described in detail in Chapter 4. The germinal region and the ducts that connect it with the exterior are continuous. The shedding of gametes into the pseudocoel would be difficult in the presence of a high turgor pressure. The reproductive system possesses its own musculature and is capable of peristaltic movements. This may be necessary for the movement of eggs to the exterior against the internal turgor pressure.

Conclusions

Nematodes provide one of the most striking examples of the relationship between structure and function. Many aspects of their functional organisation are governed by the necessity for a high turgor pressure which provides the antagonistic action to the longitudinal muscles. This has limited structural variation in many of the organ systems.

Despite this relative constancy of body form, the functional organisation of nematodes is very successful and has enabled them to exploit a wide variety of ecological niches.

3 MOVEMENT AND CO-ORDINATION

The Musculature

The somatic muscle cells are grouped in rows between the hypodermal cords, with their long axes parallel to that of the body (Figure 1.2). The musculature is classified according to the number of rows and the shape of the muscle cells (Bird, 1971); these can vary in different parts of the same nematode. The contractile region of the muscle cell may be wide and shallow, lying adjacent to the hypodermis (platymyarian), the non-contractile region may bulge into the pseudocoel (coelomyarian) or be partially or completely enclosed by the contractile region (circomyarian). The number of rows of muscle cells between the hypodermal cords also varies. There may be none or only two (holomyarian), two to five rows (meromyarian) or a large number of rows in each sector (polymyarian). The platymyarian, meromyarian pattern is the most common arrangement and may be the basic type from which more complex arrangements have evolved (Bird, 1971). The muscle cells are divided into dorsal and ventral fields by the lateral hypodermal cords, and further into four by the dorsal and ventral hypodermal cords. This division into functional fields is important in determining the pattern of nematode locomotion (see below).

The muscle cells are unusual in that they are divided into a contractile region, consisting of regularly arranged thick and thin myofilaments, and a non-contractile region, containing the nucleus, mitochondria, endoplasmic reticulum, ribosomes, glycogen and lipid droplets (Figure 3.1; Plate 3.1). The arrangement of the myofilaments forms bands which can be observed by polarising microscopy (Zengel and Epstein, 1980). The H band consists of thick myofilaments only, the A band of overlapping thick and thin myofilaments, and the I band of thin myofilaments only (Figure 3.2). This arrangement gives the muscles their striated appearance. They differ from vertebrate striated muscle in that the ends of the thick myofilaments are staggered. Nematode muscle is therefore called obliquely striated muscle. The sarcomeres or muscle blocks are often bounded by dense bodies which form the Z band and may have an associated sarcoplasmic reticulum.

Figure 3.1: Diagram of Platymyarian-type Muscle Cell. Nematode muscle cells are divided into a contractile and a non-contractile region. The contractile region consists of thick (TH) and thin (TN) myofilaments and is divided into blocks by the dense material of the Z-band. The non-contractile region consists of cytoplasm containing the nucleus (N), mitochondria (M) and lipid droplets (L). The muscle cell is connected to the cuticle by fibres (F) in the hypodermis. Original

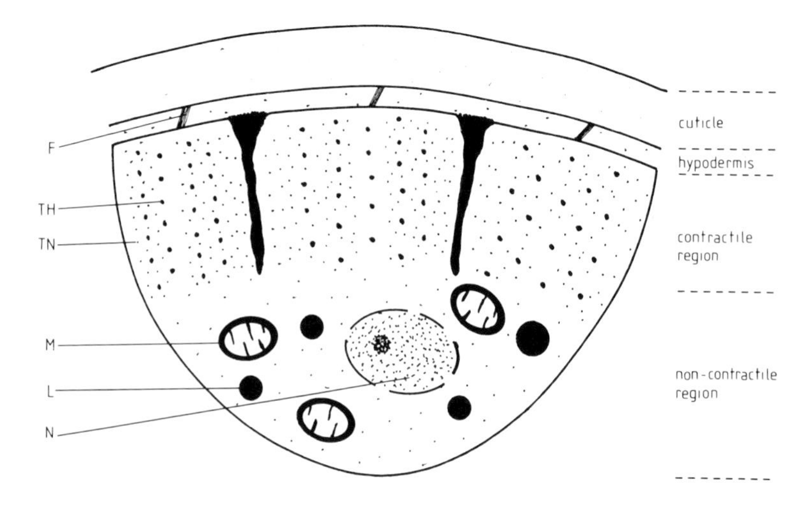

Plate 3.1: Transverse Section through the Contractile Region of the Muscle Cells of *Ditylenchus dipsaci* 4th-stage Juveniles. The contractile region consists of thick and thin myofilaments in an arrangement typical of nematode obliquely striated muscle. × 130 000. Original

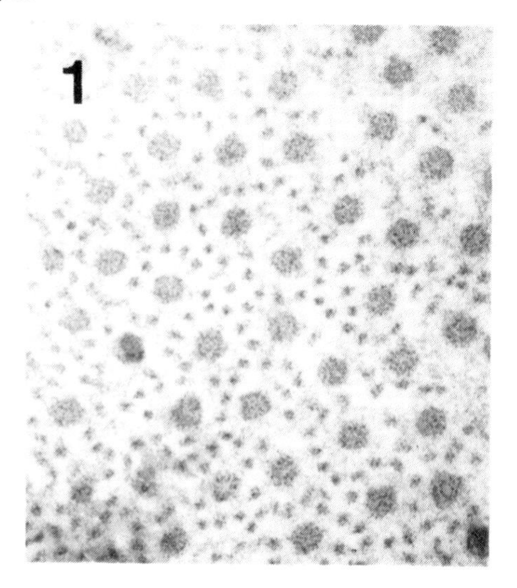

Plate 3.2: Transverse Section through the Ventral Nerve Cord of *Caenorhabditis elegans*. The nerve cord lies adjacent to the ventral hypodermal cord (H) and they are both enclosed by a basement membrane (BM). The ventral nerve cord contains the nuclei of the motoneurons (MNN), which innervate the body musculature, and various classes of interneurons (α, β), axons and dendrites (A, B, DD, VD). × 13350

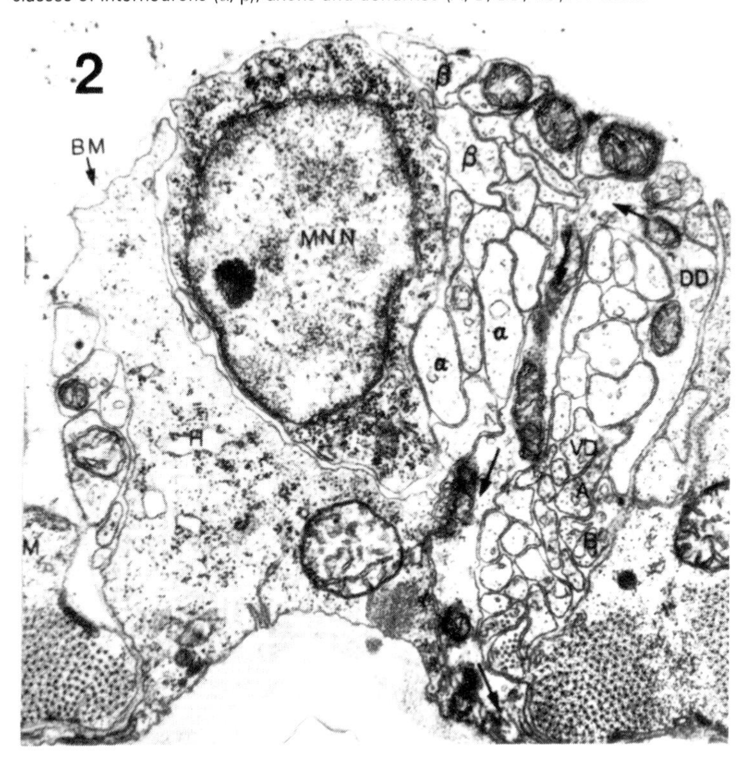

Reproduced from White *et al.* (1976), with permission

A number of contractile proteins have been identified in nematodes, including myosin, paramyosin, actin and tropomyosin (Zengel and Epstein, 1980). The myosin is similar to that of rabbit skeletal muscle and consists of two distinct molecular species. The actin is a unique molecular structure but has similar physical and chemical properties to those of other muscle actins. Nematode muscle thus has all the components necessary for a classic 'sliding filament' mechanism of muscle contraction (Huxley, 1953a, b), in which the actin-containing thin myofilaments slide past the myosin-containing thick myofilaments by the making and breaking of contacts between them (Figure 3.3). In nematodes, as in verte-

Figure 3.2: Schematic Diagram of the Obliquely Striated Muscles of Nematodes (Platymyarian). The thick and thin myofilaments and the dense bodies are arranged to form I, A, H and Z bands equivalent to those found in vertebrate striated muscle. Redrawn from Zengel and Epstein (1980)

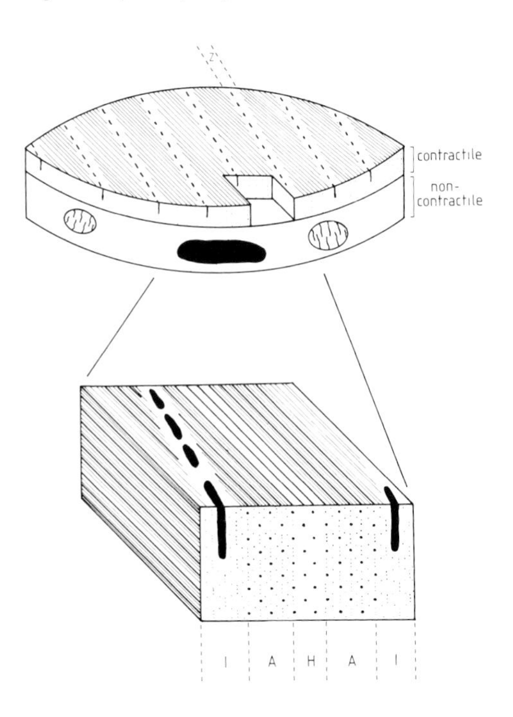

brates, this process is dependent on ATP and is regulated by calcium (Zengel and Epstein, 1980). Obliquely striated muscle can contract both by altering the interdigitation of the myofilaments and by altering the angle of shear of the whole system. They are thus supercontractile and are able to contract more than an equivalent length of cross-striated muscle (Inglis, 1983c).

The Nervous System

Structure

Our knowledge of the anatomy of the nematode nervous system has, until recently, been based largely on work done earlier this century on the large ascarids. This has recently been supplemented by studies on a small free-living nematode *Caenorhabditis elegans*

Figure 3.3: The Huxley (1953a,b) Sliding Filament Mechanism of the Contraction of Vertebrate Striated Muscle. Contraction is by the thick and thin myofilaments sliding past one another, increasing the H band and decreasing the I and A bands. Redrawn from Zengel and Epstein (1980)

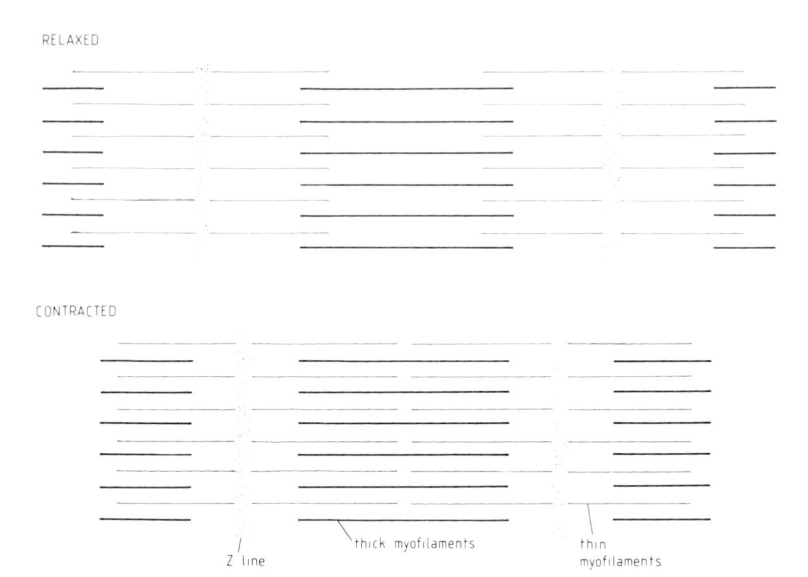

(Ward, Thomson, White and Brenner, 1975; Ware, Clark, Crossland and Russell, 1975; Albertson and Thomson, 1976; White, Southgate, Thomson and Brenner, 1976). Serial sections of entire nematodes have been cut and the nervous system mapped out with the aid of a computer reconstruction system. The basic plan of the nervous system of *C. elegans* is similar to that of *Ascaris*, again emphasising the structural uniformity of nematodes (Figure 3.4).

The anterior nervous system of *C. elegans* consists of nerves from the cephalic sense organs and motoneurons innervating the muscles of the head. The cell bodies of these neurons are situated between the two bulbs of the oesophagus to form the nerve ring. The main concentration of nerve cells is in the nerve ring, which may act as the main centre for the co-ordination of sensory information and motor activity. Most of the nerve processes leave the nerve ring on the ventral side to form the ventral nerve cord. This is the main nerve tract of the animal and lies adjacent to the ventral hypodermal cord. The oesophagus possesses its own

Figure 3.4: Basic Organisation of the Nematode Nervous System. The main concentration of neurons is in the nerve ring at the base of the oesophagus. This sends motoneurons to the musculature of the head and receives sensory neurons from the cephalic sensory receptors. The neurons forming the ventral nerve cord (VNC) leave the nerve ring on the ventral side. Neurons pass round the body at intervals as commissures and form the dorsal nerve cord (DNC). The posterior nerve ring consists of rectal ganglia and commissures and has motoneurons to the tail musculature and sensory neurons to the posterior sense organs. Original, based on the descriptions of White *et al.* (1976) and other sources

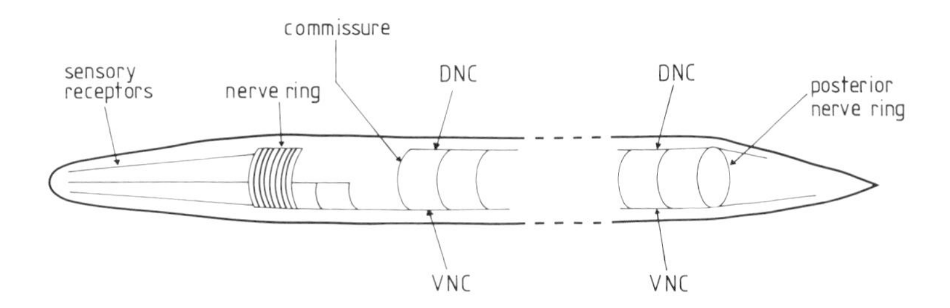

nervous system which is virtually self-contained (Albertson and Thomson, 1976).

The ventral nerve cord contains interneurons, some of which connect with ganglia in the tail and others which synapse with a series of motoneurons innervating the body musculature (Plate 3.2). Some of these motoneurons send out processes which run around the outside of the nematode as a commissure and turn to lie alongside the dorsal hypodermal cord, forming the dorsal nerve cord (Figure 3.4). The dorsal nerve cord is thus much simpler than the ventral cord, consisting mainly of motoneurons with fewer interneurons. A similar arrangement has been described in *Ascaris* (Stretton, Fishpool, Southgate, Donmoyer, Walrond, Moses and Kass, 1978).

At the posterior end, the ventral and dorsal cords end in ganglia and are connected by a series of ganglia and commissures, sometimes called the posterior nerve ring. Ganglia in this region connect with sense organs in the tail, these being far more numerous in males. In addition to the main nerve system a peripheral nervous system connecting the body setae has been demonstrated in some marine species (Bird, 1971).

The nervous systems of most species of nematode follow a similar pattern. There may be up to twelve longitudinal nerves which are connected circumferentially by commissures (Bird, 1971).

Neuromuscular Connections

The neuromuscular connections of nematodes are unusual in that rather than the motoneuron sending out a process that connects with the muscle, the muscle sends out a process (the muscle arm) that synapses with the motoneurons in the nerve cord (Figure 3.5). This has been demonstrated in *Ascaris* (Rosenbluth, 1965) and *C. elegans* (White *et al.* 1976). This arrangement has also been found in the trunk musculature of the lancelet and the tube-foot ampulla muscles of a starfish (Jarman, 1976). The membranes of the muscle cells are in close contact and held together in places by gap junctions. There is some evidence that adjacent muscle cells are electrically coupled.

Neurochemistry

Nematodes produce a number of substances which are known to act as neurotransmitters in other metazoa. These include adrenalin, noradrenalin, dopamine, dopa, 5-hydroxytryptamine (serotonin), γ-aminobutyric acid, propionylcholine and acetylcholine (Jarman, 1976; Willett, 1980).

The neuromuscular junctions are cholinergic synapses. Drugs that affect the action of acetylcholine and inhibitors that block acetylcholinesterase produce, as expected, excitatory responses in *Ascaris.* There is also histochemical evidence for the presence of acetylcholine (Lee, 1962). Acetylcholine, therefore, appears to be the main excitatory neurotransmitter in nematodes. The main inhibitory transmitter is thought to be γ-aminobutyric acid (Willett, 1980). Both of these neurotransmitters affect the state of polarisation of the muscle cell membrane. Most of the nematode

Figure 3.5: The Neuromuscular Connections of Nematodes. The muscle cell (m) sends out processes, the muscle arms (ma), which synapse with the ventral nerve cord (vn). c, cuticle; h, hypodermis. Original

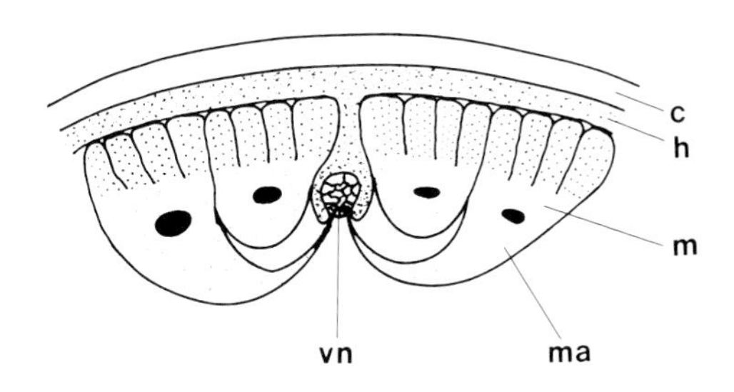

nervous system appears to be cholinergic, but other neurotransmitters have a restricted distribution and produce specific effects. Serotonin, 5-hydroxytryptophan and adrenalin stimulate activity in the vulva and vagina of a number of species (Croll, 1975a). Other putative neurotransmitters appear to be involved in neurosecretion.

Neurosecretion

A number of neurosecretory cells have been identified in nematodes on the basis of their histochemistry and ultrastructure. These include cells in the nerve ring, dorsal ganglia, ventral ganglia, ventral nerve cord and several of the sense organs. Neurosecretion may be involved in the control of moulting (Willett, 1980).

Neuromuscular Co-ordination

Because of the difficulty of inserting electrodes in small nematodes, electrophysiological studies of nematode nervous systems have largely been confined to *Ascaris*. The ventral nerve cord of *Ascaris* consists of six interneurons and seven motoneurons. Some of the motoneurons have processes which pass around the circumference of the animal as commissures to form the dorsal nerve cord. By stimulation with a suction electrode after selective surgery on the commissures, Stretton *et al.* (1978) have demonstrated that the dorsal musculature and the ventral nerve cord are electrically coupled.

Stimulation of the ventral nerve cord produces contraction of the dorsal musculature. Some of the motoneurons are excitatory and others inhibitory. If all the commissures are cut, stimulation of the ventral nerve cord has no effect on the dorsal muscles. One class of motoneurons in the dorsal nerve cord produces an inhibitory effect on the ventral musculature.

The circuitry thus exists for a reciprocal inhibition between the ventral and dorsal nerve cords. This may ensure that the dorsal muscles relax when the ventral muscles contract and vice versa — an essential requirement for the sinusoidal locomotion of nematodes.

The interneurons form continuous fibres which synapse with the motoneurons. There is little physiological information on the function of these neurons, but it is reasonable to suggest from their structural relations that they co-ordinate the activity of the motoneurons (Stretton *et al.* 1978).

A model of the known circuitry of the nervous system can be constructed which shows how it may function as a neuronal oscillator (Figure 3.6). A series of oscillators along the body could be linked in such a way that their activity would result in the propagation of waves along the body (Johnson and Stretton, 1980). Some authors, however, think that the rhythm is generated myogenically (Jarman, 1976).

Sense Organs

Nematodes show a degree of cephalisation, with the sense organs concentrated at the head end. In normal locomotion this is the part that contacts the environment first. When stationary, nematodes often exhibit head-waving or searching movements, which may represent a sampling of the immediate environment by the cephalic sense organs.

The function of nematode sense organs has largely been assigned on the basis of their structure. There is little physiological

Figure 3.6: Diagram of the Neuronal Circuitry that Has Been Demonstrated in *Ascaris*. There are connections between the dorsal excitatory (DE) and dorsal inhibitory (DI) neurons and between the ventral excitatory (VE) and ventral inhibitory (VI) neurons. Open triangles represent excitatory synapses and closed triangles inhibitory synapses. Excitatory connections VE to DI and DE to VI allow a reciprocal inhibition between the dorsal and ventral nerves and thus provide the basis for a neuronal oscillator. Synapses on to the excitatory neurons at the top of the diagram are from interneurons. Redrawn from Johnson and Stretton (1980)

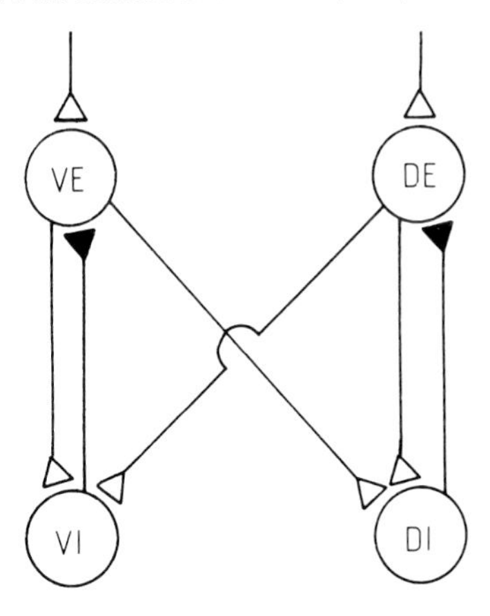

evidence of the function of any of the sense organs, and no one has succeeded in recording electrical impulses from any of these structures. Most of the sense organs are peripheral and include cuticle as part of their structure. Sense organs possessing pores which are open to the exterior are considered to be chemosensory. Where there is no pore present this does not preclude a chemosensory fuction, but cuticular modifications often suggest that they are mechanosensory (Wright, 1980). This interpretation is given some credibility by the similarity of some nematode sense organs to those of arthropods which also incorporate cuticle in their structure and on which more physiological evidence is available.

The function of sense organs has also been investigated using behavioural mutants of *C. elegans.* Mutants that show abnormal responses to chemical stimuli also show gross disruption of their presumed chemosensory organs (Ward, 1976).

Basic Organisation

The cuticular sense organs of nematodes have a very similar organisation (Wright, 1980). They consist of cuticle, a sensory dendrite or dendrites, and two non-neuronal cells, the sheath cell and the socket cell (Figure 3.7). The neuronal component consists of one or more sensory processes of the sensory neurons. A highly modified cilium connects with the end of the sensory dendrite. Wright (1980) prefers to call this a dendritic process rather than a cilium, as its structure is often very different from that of motile cilia or even some sensory cilia in other animals. The dendritic process contains microtubules which may be grouped into a circle of doublets.

The dendritic process is overlain by cuticle or, in chemoreceptors, by a cuticle-lined pore. In mechanoreceptors the cuticle is modified as papillae or setae which ensure the transfer of mechanical deformation to the receptor.

The sheath and socket cells are non-neuronal cells which may have partly a supporting function as they encircle the neuronal components of the sense organ. The tip of the socket cell is inserted into the hypodermis and is responsible for secreting the specialised cuticular component of the sense organ, in particular the cuticular 'socket' into which the dendritic process fits. The sheath cell surrounds and may have junctional complexes with the receptor region of the dendrites. It may thus regulate the environment of the sensory receptor.

Figure 3.7: The Basic Structure of Nematode Cuticular Sense Organs

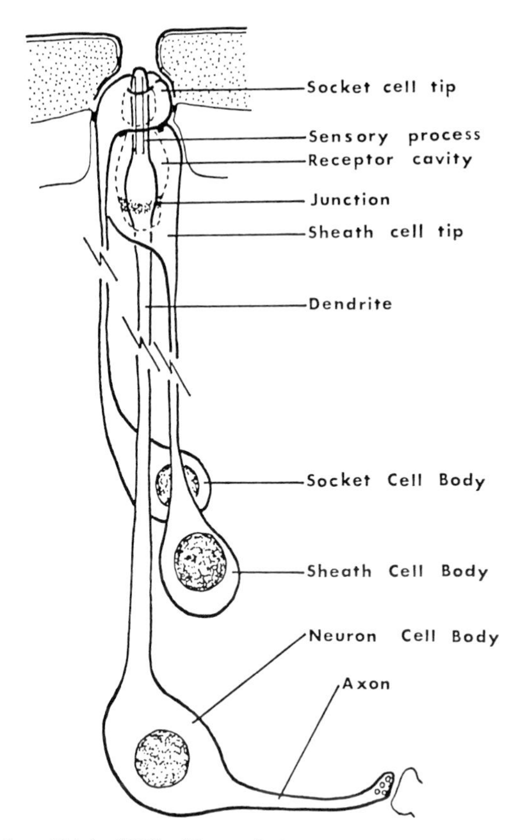

Reproduced from Wright (1983) with permission

Cephalic Sense Organs

Papillae. The cephalic papillae are arranged in concentric circles around the mouth. Their precise arrangement is of taxonomic importance and varies from group to group. The primitive pattern is hexaradial, being based on a mouth with six lips and reflecting the triradiate structure of the oesophagus (Figure 3.8). The papillae may possess cuticular projections or setae, and are considered to be mechanosensory; others have pores at their tips and may thus be chemosensory.

Amphids. Amphids are large cephalic sense organs situated

laterally on either side of the mouth (Figure 3.8). Their detailed structure varies considerably in different species, and they may have rather different functions (McLaren, 1976). The amphid consists of a cuticle-lined invagination of the body cuticle which is surrounded by a supporting cell (socket cell) and a large secretory cell (sheath cell) at its base. The secretory cell is often referred to as the amphidial gland. It may extend for as much as 25 per cent of the length of the nematode. The sensory dendrites consist of processes from cells originating in the nerve ring. These penetrate the cytoplasm of the amphidial gland cell before emerging into the lumen of the amphidial canal and connecting with their dendritic process (Figure 3.9). The amphid can have as many as 19 dendritic processes.

Amphids possess pores which are open to the exterior and are therefore considered to be chemosensory. Mutants of *C. elegans* with abnormal responses to chemical stimuli have structural defects in their amphidial dendrites (Ward, 1976). The amphids are apparently able to function as chemoreceptors even though their dendritic processes are bathed in secretions from the amphidial glands. A possible explanation is that the receptor modifies the secretory activity of the amphidial gland by moni-

Figure 3.8: The Basic Arrangement of Sense Organs around the Mouth in Nematodes. The mouth is surrounded by six lips (L) each of which bears an inner labial papillum (IP) and an outer labial papillum (OP). These are surrounded by two amphidial pores (A) and four cephalic papillae (CP). This pattern is modified in different species and is of great use to taxonomists. Redrawn from Crofton (1966)

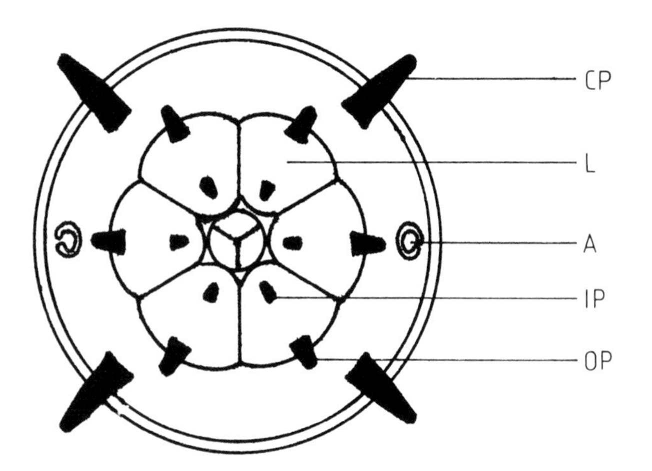

Figure 3.9: Structure of the Amphid of *Necator americanus*

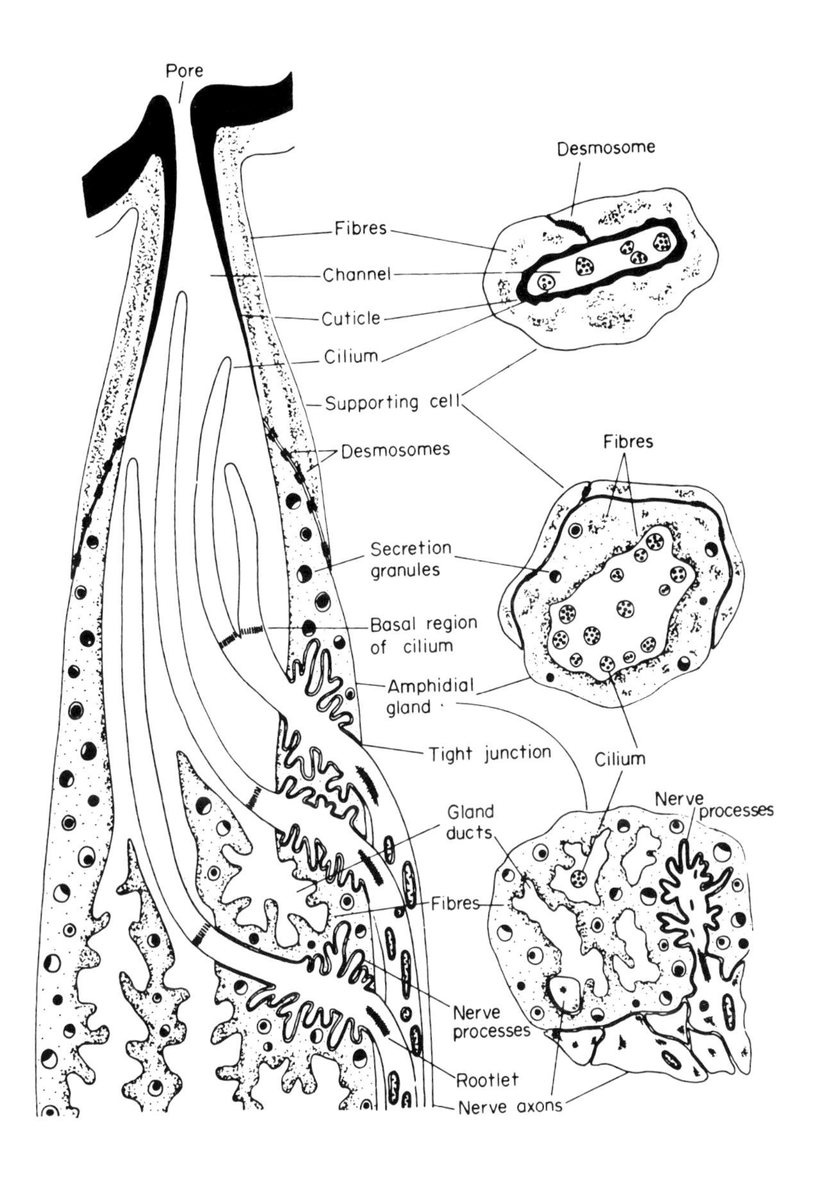

Reproduced from McLaren (1974) with permission

toring chemicals from both the external environment and gland secretions within the amphidial canal. The dendrites of the amphidial nerve penetrate the gland cell and may have processes which ramify throughout the cytoplasm.

Acetylcholinesterase has been detected in the amphidial gland secretions of a number of nematodes (McLaren, 1976). This either alters the permeability of host membranes causing them to leak nutrients or acts as a 'biochemical holdfast' by damping down the peristaltic movements of the host gut. The amphidial glands of a number of blood-feeders synthesise anticoagulants.

Posterior Sense Organs

Phasmids. The phasmids have attained a peculiar significance because nematodes have often been divided into two groups, the Adenophorea (Aphasmidia) and Secernentia (Phasmidia), on the absence or presence of these structures (although there are other differences — see Chapter 1). The phasmids are paired lateral sense organs which open via pores on the tail. They may be chemosensory, and the presence of acetylcholinesterase has been demonstrated in the phasmids of *Dipetalonema vitae* (Wright, 1980).

Male Sense Organs. Males possess a number of sensory structures which are involved in the control of copulation. Caudal papillae are found on the ventral surface of the tails of male nematodes. The cuticle of the tail may be extended to form a bursa which holds the female during copulation. The bursal rays contain sensory dendrites connecting with sensory papillae at the periphery of the bursa.

Males also possess spicules, cuticular accessory reproductive structures which are protruded through the cloaca during copulation. The spicules possess a number of sensory structures which have the features of both mechano- and chemoreceptors (Wright, 1980).

Other Sense Organs

These include the deirids which are situated on either side of the body posterior to the nerve ring, body pores in dorylaimid nematodes and elongate body setae of freshwater and marine nematodes. In addition there are a number of internal receptors which do not have a cuticular component. These include dendrites in the mouth, oesophagus, intestinal-rectal valve and in the body wall (Wright, 1980). Some of these may have a proprioceptive function. A number of nematodes are photosensitive and some possess pigment spots which form part of a photoreceptor (Wright, 1980).

The hemizonid consists of nervous tissue underlying an area of modified cuticle in the region of the nerve ring. It is prominent in trichostrongyle infective juveniles and may be involved in the exsheathment process. Wright (1980) considers it to be a large nerve commissure with no sensory function.

Locomotion

Undulatory propulsion is the basic pattern of locomotion in nematodes. This is the kind of locomotion found in snakes, eels and other animals (Gray, 1968). In snakes, bending is produced by muscular contraction acting against the rigid vertebral column. The turgor pressure system of nematodes acts in a similar fashion, by providing a hydrostatic skeleton which allows the body to bend when the dorsal or ventral muscles contract. As nematodes lie on their sides, bending couples act against the substrate; resulting in forward locomotion from backward waves. As we have seen (Chapter 2), the turgor pressure system also provides the antagonistic action to the longitudinal muscles.

The body form of nematodes is well adapted to undulatory propulsion. They are elongate cylinders, pointed at either end. The head or tail can easily find a path between particles, which the rest of the body follows by a travelling sine wave. The streamlined body shape helps the nematode to move through viscous media.

The functional division of the musculature into dorsal and ventral fields and the way in which the muscles are innervated are also suited to undulatory propulsion. Contraction of one set of muscles produces extension of the muscles on the other side of the body. Sine waves are formed by waves of contraction passing along the dorsal and ventral musculature out of phase with one another (with a phase difference of 180°) (Figure 3.10). Such a wave generates movement via the tangential forces acting upon the substrate or medium through which the nematode is moving (Figure 3.11).

The degree of friction between the nematode and the substrate

Figure 3.10: The State of Contraction of the Body Wall Muscles (Diagrammatic Longitudinal Section) during Undulatory Propulsion. A zone of contraction in the longitudinal muscles is situated opposite a zone of relaxation on the other side of the body. Redrawn from Crofton (1966)

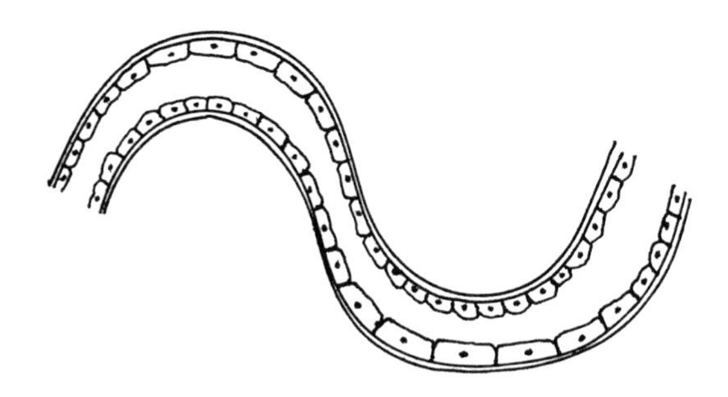

determines the efficiency of locomotion. If friction is decreased to a point where it can no longer oppose the force generated by the sine wave (see Figure 3.11c, $F \cos \theta < N \sin \theta$), for example in a medium of low viscosity, the nematode compensates by decreasing θ; that is by forming waves of smaller amplitude and greater wavelength. Locomotion is at its most efficient when the nematode travels forwards at the same speed as the waves pass backwards. Nematodes often move in a film of water, such as between soil particles or over the surface of plants. Here the friction and the normal reaction forces are dependent on the surface tension of the water film acting upon the nematode (Figure 3.12). In a thin film the friction may be sufficient to produce optimum locomotion but in a thicker film or layer the nematode can only move by crawling over the substrate or by swimming.

When swimming, the friction forces are very low although resistance may be increased by enlarged lateral alae. The nematode develops a high propulsive thrust by increasing wave frequency. Only a few species can achieve this. Increasing wave frequency results in a corresponding increase in wavelength, and most swimming nematodes have only a single wave which exhibits two stationary nodes (Figure 3.13).

Whereas most species move by simple undulatory propulsion, some can move by forming helical waves which leave a corkscrew-like path when the animal is swimming through agar. It is difficult to explain this with our current knowledge of nematode neurophysiology. The body may have a basic helical organisation with the organs rotating around the longitudinal axis (Crofton, 1971a)

Figure 3.11: (a) A Nematode Moving along a Sinusoidal Path between Soil Particles. Each half wave acts as a propulsive unit which moves the nematode forwards. (b) At Each Point of Contact with a Soil Particle the Forces Consist of a Normal Reaction Force (N_1-N_3) between the nematode and the particle and a friction force (F_1-F_3) acting along the nematode. (c) These Forces Can Be Resolved in the Direction the Nematode is Moving and at Right Angles to the Axis of Propulsion

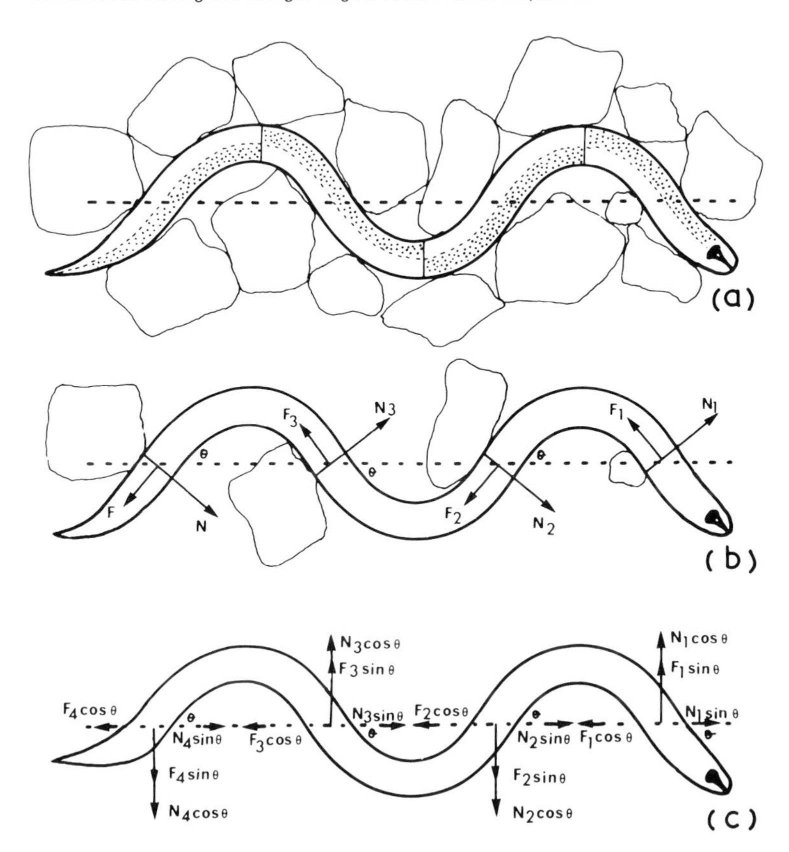

Reproduced from Wallace (1968) with permission

but there is little evidence of this: the lateral alae are straight and do not rotate around the body (Wharton, 1982a). Alternatively the left and right dorsal/ventral muscle fields may contract out of phase with one another. The four muscle fields contracting in sequence would produce a helical configuration (Figure 3.14).

Only a few species exhibit forms of locomotion other than undulatory propulsion (Adams and Tyler, 1980). The infective

Figure 3.12: In a Film of Water, Locomotion is Dependent on the Surface Tension Forces Acting upon the Nematode. This diagram shows these forces in a thick film (a) and a thin film (b), and in a thin film when the nematode is in motion (c). A thin film of water is displaced when the nematode moves (c). *T* surface tension force, *N* normal reaction force between the nematode and the substrate, *R* resultant of the surface tension forces in (c), *F* friction force, μ the coefficient of friction, *P* the nematode's propulsive force

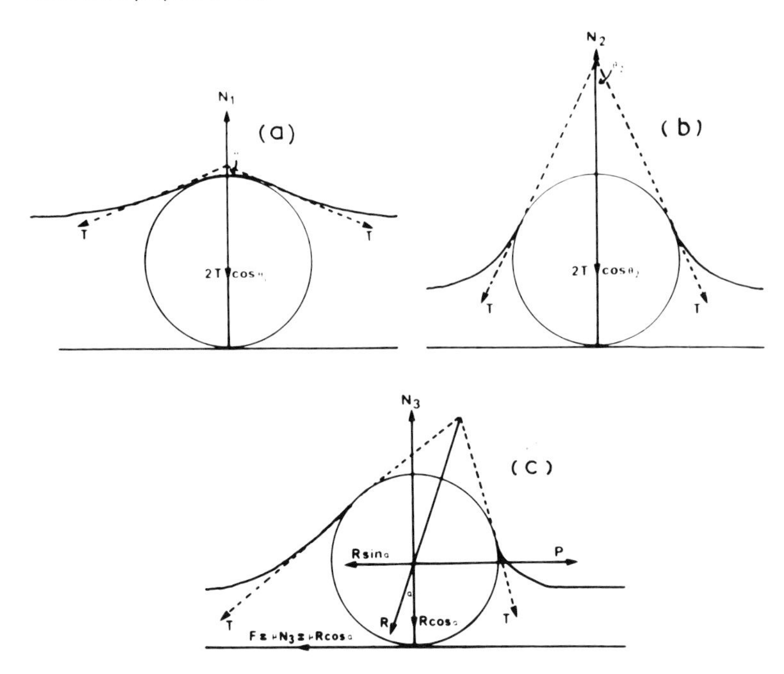

Reproduced from Wallace (1968) with permission

Figure 3.13: When Swimming a Nematode Exhibits a Single Wave, with Two Stationary Nodes. Original

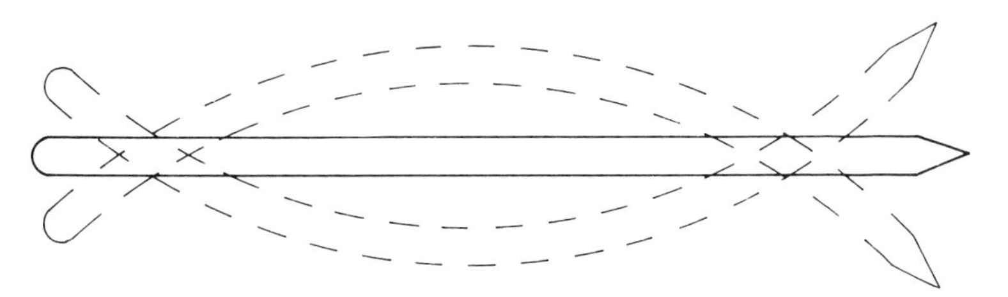

Figure 3.14: Undulatory Locomotion and the State of Muscular Contraction in (a) a Single Plane, and (b) when Spiralling. When spiralling, the left dorsal, right dorsal, right ventral, left ventral muscles contract in sequence. Black circles represent contracted muscle, white circles relaxed muscle and stippled circles partially contracted muscle

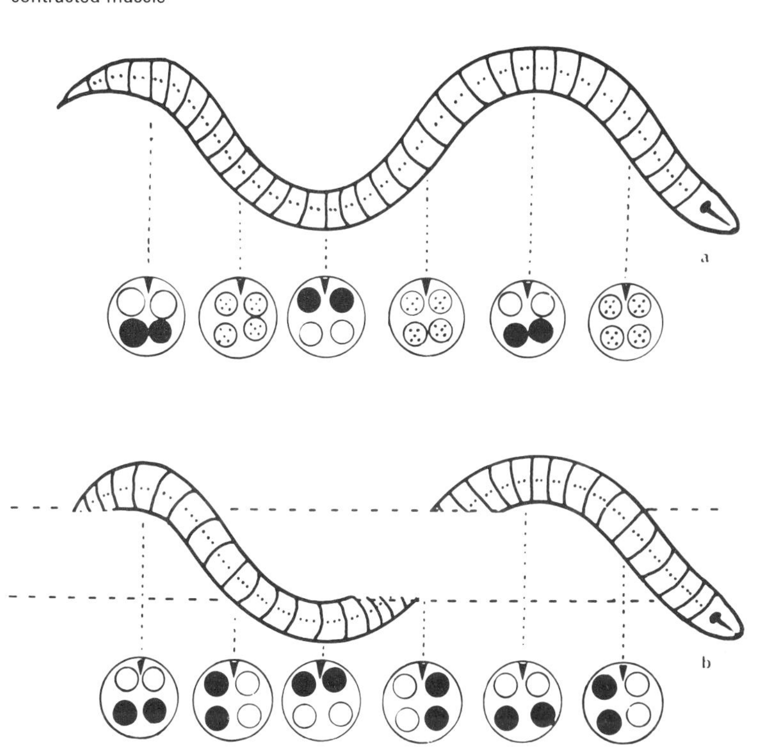

Reproduced from Wallace (1968) with permission

juvenile of *Neoaplectana carpocapsae* can mount projections on the substrate and bend its body until the head makes contact with the tail. The head and tail are held together by surface tension forces until the nematode develops enough power to overcome this force and leaps through the air. Some species move by the alternate attachment of head and tail to the substrate (looping) or by the attachment of the tail alone (hopping); in conjunction with the bending and stretching of the rest of the body. Anterior and posterior bristles and the caudal glands are used for attachment to the substrate (Adams and Tyler, 1980). *Circonemoides* sp. has overlapping cuticular annulations which form backward-facing projections. A wave of contraction of both the dorsal and ventral musculature has been observed to pass backwards along the body. This, in conjunction with the cuticular projections which prevent slippage, would result in forward locomotion — in a similar fashion to that of an earthworm.

It is difficult to interpret the aberrant patterns of locomotion observed in these species given the sparsity of our knowledge of their structure and physiology. They do, however, show that a few species have been able to modify the basic functional organisation of nematodes and must possess a rather different neuromuscular physiology than that involved in undulatory propulsion.

Behaviour

A wide variety of behavioural activities are observed in nematodes (Table 3.1). These can be divided into basic behaviours, which are widely found among nematodes, specialised behaviours, which are found in a restricted range of species, and behavioural sequences, which consist of a series of basic and specialised behaviours that result in a particular activity. Some behaviours are initiated exogenously by a stimulus detected by the sense organs. Others appear to arise spontaneously and are thought to be under endogenous control (Croll and Sukhdeo, 1981). Endogenous control may involve myogenic pacemaker activity, the monitoring of muscle tone by proprioceptors, and be dependent upon the turgor pressure of the pseudocoel. It should be noted that it is the stimu-

Table 3.1: Classification of Nematode Behavioural Activities

Basic behaviours	Specialised behaviours	Behavioural sequences
Inactivity	Air breathing	Feeding
Refractory state	Coiling	Prey capture
Head waving	Caudal gland attachment	Exsheathment
Oesophageal pumping	Nictating	Hatching
Defecation	Stylet thrusting	Copulation
Bursal flapping	Leaping	Oviposition
Forward waves	Swarming	Migration and dispersal
Backward waves	Spiralling	Penetration
Shock reactions	Anabiosis	Site selection
Moulting		Orientation to gradients
Movement in the egg		

Taken from Croll (1972), Wharton (1981) and Croll and Matthews (1977)

lus initiating the behaviour that comes from outside or within the animal, rather than the behavioural activity being classified as exogenous or endogenous. Behaviours that are often considered to be endogenous (oesophageal pumping, defecation, bursal flapping) may form part of a behavioural sequence which is initiated by an exogenous stimulus (feeding, copulation).

There appears to be a distinct hierarchy in behavioural activities. Rapid body movement ceases during feeding. Behavioural sequences (moulting, copulation, etc.) once initiated will result in a series of activities which will not be interrupted until the sequence is complete.

The type of behavioural activity initiated by an exogenous stimulus varies, depending upon the nature of the stimulus. The infective juveniles of trichostrongyle nematodes are non-feeding and are inactive for most of the time. Mechanical disturbance initiates a burst of locomotory activity (Figure 3.15). An increase in temperature, however, causes the juveniles to coil (Wharton, 1981). Croll and Sukhdeo (1981) recognise a distinct sequence which results in locomotory or feeding activity in many nematodes (Figure 3.16).

When a nematode encounters an obstacle, the wave direction reverses (backward to forward waves) and the worm swims back-

Figure 3.15: The Infective Juveniles of *Trichostrongylus colubriformis* Respond to Mechanical Disturbance with a Burst of Activity. This declines to a low level of endogenous activity. Redrawn from Wharton (1981)

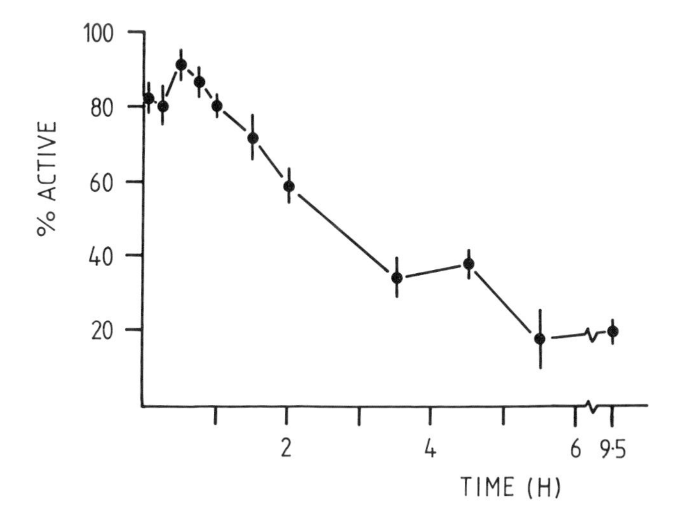

Figure 3.16: Relationship between the Basic Behavioural Activities Exhibited by Nematodes. Redrawn from Croll and Sukhdeo (1981)

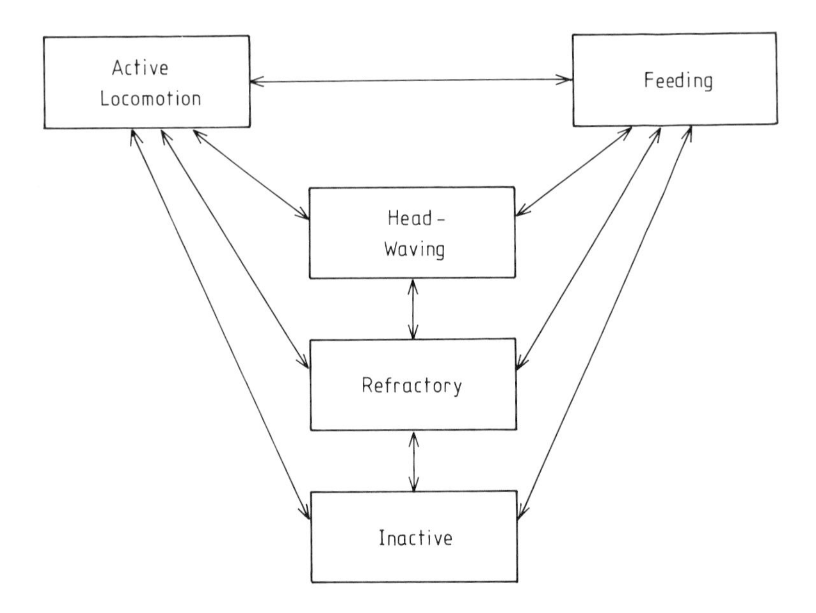

wards for a short distance before resuming forward locomotion in a different direction. The tracks made by nematodes moving on the surface of agar indicate that nematodes also undergo spontaneous reversals which result in a change in the direction of forward locomotion. Croll (1975b) suggested that these, and other behavioural interactions, could be explained by a 'two-generator hypothesis', in which two myogenic pacemakers control body movements. One is situated in the circum-oesophageal ganglion or nerve ring and the other in the posterior ganglia. The two generators initiate waves in opposite directions, the anterior generator initiating backward waves, resulting in forward locomotion, and the posterior generator initiating forward waves, resulting in backward locomotion. The anterior generator is the dominant control centre, resulting in several minutes' activity. If both generators operate together, it may result in a refractory state in which the nematode is insensitive to further stimuli. Though a great deal of nematode behavioural activity can be interpreted in terms of this hypothesis, there is, as yet, little neurophysiological evidence that supports the existence of these two generators.

Nematodes exhibit orientation responses to a variety of physical and chemical stimuli (Croll and Matthews, 1977). These responses are important for avoiding adverse conditions, food capture, finding a mate, and host location. Chemosensory responses have been the most thoroughly investigated. A variety of substances act as attractants to *C. elegans* and rather fewer as repellants (Table 3.2). Responses to attractant substances may be important in the location of bacterial food, although there is no response to bacterial filtrate (Ward, 1978). A number of species are known to produce sex attractants (Green, 1980). Carbon dioxide has been shown to act as an attractant to the infective juveniles of some plant and animal parasitic nematodes and may help in host location in these species (Croll and Matthews, 1977).

The ability of nematodes to move up a concentration gradient implies that they are able to detect changes in the concentration of the chemical concerned. Detection of a concentration gradient may involve sampling by two receptors separated in space or by a single receptor sampling at different points in time (Ward, 1978). The separation of the two amphids and their relative positions suggest that they are unlikely to detect a concentration gradient. Sufficient separation may be provided by the simultaneous use of the amphids in the head and the phasmids in the tail. This is unlikely, however, as *C. elegans* blister mutants, which have cuticle covering the phasmids, can orientate normally, and phasmids are absent in adenophorean orders.

Sampling in time may be provided by the forward locomotion of the nematode or by the side-to-side movement of the head during locomotion. Muscle-degenerate mutants of *C. elegans* move slowly but can nevertheless orientate correctly. If forward movement

Table 3.2: Chemicals which Elicit Chemotactic Responses in *C. elegans*

Attractants	Repellants
Cl^-, SO_4^-, NO_3^-, Br^-, I^-	H^+ (acid)
Na^+, Li^+, K^+, Mg^{2+}, Ca^{2+}	D-tryptophan
cAMP, cGMP	CO_2, H_2CO_3 or HCO_3^-
OH^-	(CO_2 phosphate)
Pyridine	
HCO_3^- or CO_3^{2-} (CO_2 borate)	
CH_2COO^-	

Taken from Ward (1978)

were important, these mutants would be expected to orientate less accurately than wild types. It therefore appears that side-to-side movement of the head is important for the orientation response. Short-headed mutants, which have a shorter span of side-to-side head movement, do not orientate as rapidly as wild types (Ward, 1978). Head-waving behaviour has been interpreted as sampling the environment, and is thought to be important in the response to sex attractants (Green, 1980).

The side-to-side movement of the head of *C. elegans* is such that it might be able to detect a 3 per cent change in concentration. Ward (1978) has suggested that detection acts by the interaction between a chemical and receptor sites on chemosensory neurons in the amphid, in a similar fashion to chemoreception in bacteria. The number of classes of receptor site can be estimated by the competition between the responses to two chemicals. If the nematode can orientate correctly to a gradient of one chemical in the presence of a uniformly high concentration of another, and vice versa, the two chemicals must affect different receptor sites. Together with the analysis of behavioural mutants, this suggests that there are at least nine classes of receptor site (Ward, 1978).

If this interpretation is correct, we may expect that, as the number of receptor sites available must be limited, the response will fall off as the concentration of the chemical increases and as all the receptor sites become occupied. This can be predicted by the equation:

$$R = c \frac{\mathrm{d}A}{\mathrm{d}t} \frac{K_\mathrm{d}}{(K_\mathrm{d} + A)^2}$$

where R is the response, A is the attractant concentration, t is time, c is a constant and K_d is the dissociation coefficient of binding. The response to pyridine does show such a pattern (Figure 3.17), although the response to other chemicals may be more complicated (Ward, 1978).

Surface carbohydrates are localised around the sense organs of several species of nematode and are thought to originate as secretions of these organs. Enzymatic removal of the carbohydrate or exposure to specific carbohydrate-binding lectins interferes with the chemotactic response. Carbohydrate residues associated with the sensillae may aid the diffusion of chemotactic factors towards the receptor and perhaps act as binding sites on the receptor

Figure 3.17: The Effect of Pyridine Concentration on the Chemosensory Response of *C. elegans* to this Chemical. Redrawn from Dusenbery (1976)

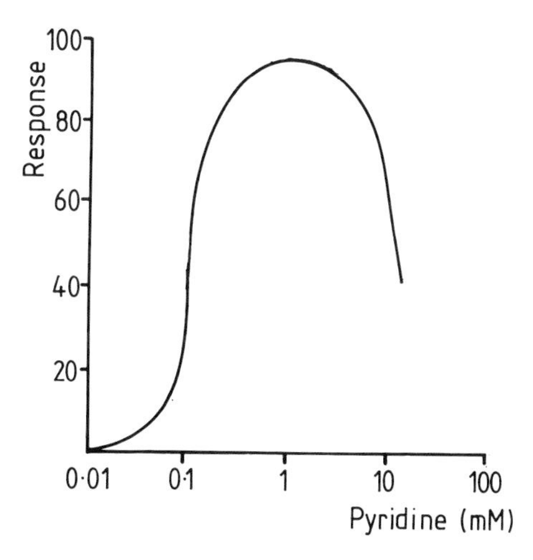

membrane. Blockage or removal of these binding sites provides a potential means of interfering with the chemotactic responses involved in host and mate location and may be useful in the control of some nematode pests (Zuckerman and Jansson, 1984).

4 REPRODUCTIVE BIOLOGY

The reproductive potential of nematodes is often very great. *Ascaris lumbricoides* can lay 200 000 eggs a day and *C. elegans* can reproduce itself in 3.5 days at 20°C (Brenner, 1974). There is great variation in fecundity and generation time between different species, reflecting the life-cycle and reproductive strategies employed.

Reproductive Mechanisms

The majority of nematodes reproduce sexually. The zygote is formed by the fusion of male and female gametes (sperm and eggs) which originate from the germ cell line by a series of mitotic and meiotic divisions. In amphimictic species egg and sperm come from separate individuals and the species thus possesses male and female sexes (dioecious or gonochoristic). Males and females differ in their reproductive systems, accessory sexual organs and often in size and general body shape. Intersexes, which have both male and female characteristics, occur in some species.

A number of free-living species and rather fewer parasitic species reproduce hermaphroditically (egg and sperm are produced by the same individual) or parthenogenetically (the egg develops without fertilisation by a male gamete). All nematodes, however, reproduce from gametes produced by the germ cell line. There is no proliferation of individuals from the somatic cell line as is found, for example, in the vegetative growth of corals or the budding of *Hydra.*

The Significance of Sexual and Unisexual Reproduction

A wide variety of reproductive mechanisms are found in nematodes. It may, therefore, be useful to consider the advantages and disadvantages of different modes of reproduction in relation to various nematode habitats.

All eggs produced by a parthenogenetic female can develop into egg-laying females. In an amphimictic species, only half the eggs can do so (the other half producing males). The parthenogen therefore has a 'two-fold advantage' in reproductive terms over an amphimictic species (Maynard Smith, 1978). A parthenogen can thus colonise a habitat much more rapidly than an amphimictic species. In addition, sexually reproducing species have to expend energy on mate location, copulation and fertilisation, reducing the relative returns from the reproductive process (Calow, 1978).

The reproductive capacity of a parthenogenetic species is, however, achieved at the expense of a reduction in genetic variability and thus in the potential for adaptation. An offspring produced by mitotic parthenogenesis is approximately genetically identical to its parent. Mutations are the only source of variability and these usually occur at quite low rates in living organisms. Sexual reproduction, however, results in genotypic and phenotypic variation by the independent assortment of chromosomes of the two partners in the zygote and the opportunity for crossovers provided by meiosis.

There are a number of theories that suggest short-term and long-term advantages which accrue to phenotypic variation and hence the maintenance of sexual reproduction (Maynard Smith, 1978; Bell, 1982). In an unpredictable physical and biological environment a sexually reproducing species is more likely to produce at least some phenotypes which can survive these changes and reproduce. A parthenogenetic clone possesses only one phenotype (unless a mutation occurs) and may become extinct if the environment changes. Changes in the biological environment of a species may be particularly important in maintaining sexual reproduction. A species must evolve continuously if it is to survive changes in its competitors, predators and pathogens (the 'Red Queen' hypothesis: you have to keep running in order to stay in the same place). Alternatively, in a physically and spatially heterogeneous environment a sexual species with varying phenotypes will be able to occupy a greater range of niches within that environment than would a single asexual clone (the 'tangled bank' hypothesis). We may therefore expect sexual reproduction to be favoured in biologically more complex and temporally and spatially variable environments, and parthenogenesis to be favoured in environments that are relatively stable and in which the species experiences little competition or predation.

The reproductive mechanisms found in nematodes conform with these hypotheses. Hermaphroditism and parthenogenesis occur in species that are able to rapidly colonise a fresh habitat. Parasitic species face variability in the physiology, behaviour and immune reactions of their hosts, as well as in the environment outside the host which affects the chances of transmission. Sexual reproduction is, therefore, favoured and most animal parasitic nematodes do reproduce amphimictically.

Amphimixis

Most nematode species possess separate males and females and reproduce amphimictically. Descriptions of the morphology and physiology of the reproductive system later in this chapter refer mainly to amphimictic species. Although sexual reproduction confers advantages of variability, the production of males results in the two-fold disadvantage in reproductive capacity compared with parthenogenetic species. Amphimictic species also need to find a mate for reproduction to proceed. This may be a problem for a parasite at low levels of transmission. When infection levels are low, the probability of pairing may fall below the threshold level at which the population can be maintained and extinction will follow (Anderson, 1982).

Nematodes possess adaptations which aid mate location and the efficiency of sperm utilisation once mating has occurred. Mate location is aided by the production of pheromones which are attractive to the opposite sex (Green, 1980). The females of most species store sperm after copulation at the junction of the uterus and the oviduct, or in a modification of this region called the seminal receptacle. This may reduce the number of mating contacts necessary. The females of some species form permanent associations with their males. The males of *Syngamus tracheae* are much smaller than the females and lie permanently attached to the vagina. The male of *Trichosomoides crassicauda* actually lives within the uterus of the female (Poinar and Hansen, 1983). In many nematodes the male is much smaller than the female. This may reduce the proportion of the parents' resources that need to be spent on the production of 'unproductive' males.

In those species that possess sex chromosomes, sex determination is by an XY or XO system. The homozygous form (XX) produces females and the heterozygous condition (XY,XO) males. Plant parasitic nematodes appear to lack distinct sex chromosomes (Goldstein, 1981). Intersexes, showing a mixture of male and female characters, occur in a number of nematode species. They are usually female in their reproductive system but possess male accessory sexual organs (copulatory bursa and spicules). In some

Meloidogyne spp., intersexes are formed as a result of nutritional deficiencies. *Caenorhabditis elegans* mutants that produce intersexes indicate that the gonads and the accessory reproductive structures are under independent genetic control. Intersex production may thus occur as a result of the interruption of development (Poinar and Hansen, 1983).

Most species possess a sex ratio of 1:1. Under adverse conditions the proportion of males may increase as a result of differential survival. This increases the chances of mating in a deteriorating environment. The environment may also have a more direct effect on sex determination. In mermithid nematodes the proportion of male parasitic juveniles within the insect host increases with the size of the infection. The sex of an invading juvenile appears to be influenced by the juveniles already present. This may be chemically mediated by a male-determining factor which is released by the parasitic juvenile (Poinar and Hansen, 1983).

Hermaphroditism

In an hermaphroditic species egg and sperm are produced by the same individual. All individuals are available for reproduction and hermaphrodites thus possess the two-fold advantage in reproductive capacity over amphimictic species. Hermaphrodites also have an increased probability of mating at low population densities. However, cross-fertilising hermaphrodites, which are common in platyhelminths, do not occur in nematodes. Self-fertilising hermaphrodites have complete independence from mating contacts. More genetic variability is achieved than in mitotic parthenogens by allowing crossovers during meiosis and the independent assortment of chromosomes, but self-fertilisation results in much lower fitness than outcrossing because the increased homozygosity allows the expression of disadvantageous recessive genes.

Hermaphroditic nematodes are protandrous; they produce sperm first followed by eggs in a single organ, the ovotestis. The sperm produced are stored until the oocytes are mature and then fertilisation occurs. By utilising a single reproductive organ for the production of both eggs and sperm, nematodes avoid the energy costs incurred by hermaphrodites which have both male and female systems (Calow, 1981).

Pseudogamy

In pseudogamous species the sperm triggers the cleavage divisions of the oocyte but plays no further part in its development and does not contribute any genetic material to the zygote. Pseudogamy is found in both amphimictic and hermaphroditic nematodes and was first described in *Rhabditis aberrans* (Nicholas, 1984). It is sometimes considered to be an intermediate stage in the development of parthenogenesis (Poinar and Hansen, 1983).

Parthenogenesis

The unfertilised oocyte of parthenogenetic nematode species develops without any intervention by a male gamete, usually resulting in the formation of diploid or polyploid females. Both mitotic and meiotic parthenogenesis occurs. In mitotic parthenogenesis (apomixis) the zygote is formed by mitotic division and thus retains the diploid chromosome number ($2n$). In meiotic parthenogenesis (automixis) the zygote is formed by a reduction division, resulting in the haploid condition (n). The diploid chromosome number is restored by the chromosomes regrouping at telophase II to form a single pronucleus (parthenogenetic strains of *Aphelenchoides avenae*), fusion between the oocyte pronucleus and the nucleus of the polar body (*Meloidogyne hapla*) or by the chromosomes doubling without a cleavage division (*Heterodera betulae*) (Poinar and Hansen, 1983). Meiotic parthenogenesis introduces a degree of variability by allowing crossovers and the independent assortment of chromosomes but, as with self-fertilising hermaphrodites, increases homozygosity and the expression of deleterious recessive genes (Calow, 1981).

Male Production by Hermaphroditic and Parthenogenetic Nematodes

The increased reproductive capacity of parthenogens and self-fertilising hermaphrodites occurs at the expense of genetic variability and hence biological adaptability. Some species can, however, increase variability by producing males spontaneously or in response to environmental conditions. *Caenorhabditis elegans* is normally hermaphroditic but produces males spontaneously in culture with a frequency of about 1 per 700 offspring. Males mate with hermaphrodites and fertilise oocytes at a high efficiency; that is, each sperm fertilises an oocyte. Male sperm fertilise oocytes in preference to hermaphrodite sperm, possibly by reducing the fertility or adherence and motility of hermaphrodite sperm stored in the reproductive tract. The reproductive efficiency of the occasional males thus allows enough outcrossing to occur to prevent genetic uniformity in the population (Ward and Carrel, 1979).

The free-living nematode *Diplenteron potohikus* exhibits facultative parthenogenesis. Isolated females reproduce parthenogenetically but will mate with males which appear in culture at a rate dependent upon the density of the worms. Male production appears to be a response to a metabolite released by the nematode (Clark, 1978).

Second-stage juveniles of *Meloidogyne* spp. develop into females when conditions are favourable, but when under stress most develop into males. Conditions that induce male formation include crowding, low host susceptibility and stress in the host plant caused by growth regulators, decapitation, infected roots and high or low temperatures (Triantaphyllou, 1971). Male production in reponse to crowding or stress may increase genetic variability in a deteriorating environment.

Heterogamy

Some rhabditid and strongylid nematodes possess complex life cycles with an hermaphroditic or parthenogenetic stage parasitic in vertebrates, and a free-living cycle where the nematode may reproduce amphimictically for a single or for several generations. In heterogamous nematode parasites of insects, the amphimictic cycle occurs within the host (Poinar and Hansen, 1983).

Heterogamy may have evolved as an adaptation to the colonisation of successive different environments. Variation is maintained by the amphimictic cycle and a successful parasitic genotype is rapidly reproduced by parthenogenesis or hermaphroditism, obviating the need to locate a mate within the host.

The Evolution of Reproductive Mechanisms in Nematodes

Hermaphroditism is usually considered to be a primitive condition in animals and to have given rise to amphimictic species. In nematodes, however, it is thought that hermaphrodites arose from amphimictic species (Triantaphyllou and Hirschmann, 1964). Intermediate stages in the development of hermaphroditism from amphimixis can be demonstrated in rhabditids. *Rhabditis vigueri* possesses normal males (4.5 per cent of the population), normal females (10-20 per cent) and hermaphrodites. In other rhabditids normal females are absent and normal males become increasingly rare. Hermaphroditism occurs among several groups of nematodes and appears to have arisen independently several times.

Parthenogenesis is also thought to have evolved from amphimixis in nematodes. The phenomenon of pseudogamy may represent a stage in its evolution. Parthenogenesis is rare in animal parasitic nematodes but the oxyurid nematode *Gyrinicola batrachiensis*, which parasitises amphibians, exhibits haplodiploidy; in which males develop from unfertilised eggs and females from fertilised eggs (Adamson, 1981). This may also represent a stage in the development of parthenogenesis.

Parthenogenesis appears to have evolved several times among free-living and plant parasitic nematodes. Even closely related species may exhibit amphimixis and parthenogenesis respectively. *Aphelenchoides avenae* possesses both amphimictic and parthenogenetic strains (Poinar and Hansen, 1983). In animal parasitic nematodes, parthenogenesis occurs mainly as part of a heterogamous cycle.

The variety of reproductive phenomena found in nematodes and the ease of culture and mutant isolation in free-living nematodes such as *Caenorhabditis elegans* make nematodes attractive models for the study of the evolution of reproductive mechanisms in animals.

The Structure of the Reproductive System

The Male Reproductive System

Most nematodes possess a single male gonad (monorchic) although some genera possess two (diorchic). The male gonad is divided into the testis, where sperm are produced, the seminal vesicle and the vas deferens (Figure 4.1). In some species there is a vas efferens between the testis and the seminal vesicle. In most species germ cell proliferation occurs at the blind end of the tubular gonad (telogonic) but in trichurids and dioctophymids proliferation occurs throughout the length of the testis (hologonic). In hologonic testes, proliferation and growth occur throughout the length of the testis, although the actual germinal zone may be restricted to one side. Germ cell proliferation varies in different parts of the testis, producing a cycle of proliferation which ensures the continuity of sperm production. This arrangement has been

Figure 4.1: The Structure of the Male Reproductive System of *Rhabditis strongyloides* (A), *Aspiculuris tetraptera* (B), and *Trichuris suis* (C). SV, seminal vesicle; T, testis; VD, vas deferens; VE, vas efferens. Redrawn from Chitwood and Chitwood (1974) and Anya (1976)

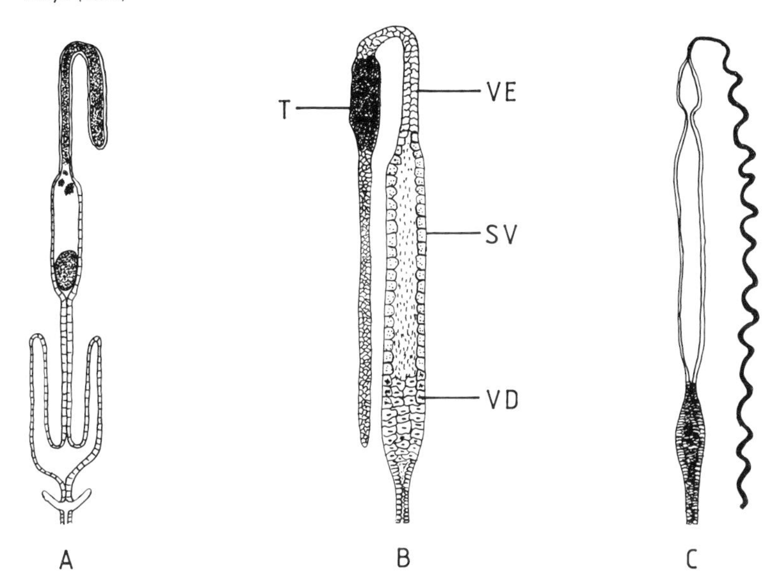

compared to the seminiferous tubules of the vertebrate testis (Jenkins, Larkman and Funnell, 1979). Telogonic testes are divided into germinal and growth zones.

Males also possess accessory reproductive structures which are involved in copulation. These are modifications of the cuticle and are thus collagenous and not chitinous, as has sometimes been suggested.

The Female Reproductive System

The female reproductive system is also tubular and is classified according to whether there are one or two gonads (monodelphic or didelphic) and according to the direction of the gonads in relation to their opening at the vagina (Bird, 1971). The female gonad is divided into the ovary (which again may be telogonic or hologonic), oviduct, seminal receptacle (where, if present, sperm are stored after copulation), uterus and vagina. The mature oocytes are released into the oviduct and are fertilised by sperm stored in the seminal receptacle or the upper part of the uterus. Fertilisation initiates egg-shell formation, which continues as the eggs pass down the uterus. Mature eggs are stored in the lower part of the uterus (vagina uterina) before laying via the vagina (vagina vera) (Figure 4.2).

The uterus possesses a layer of muscle cells which move the eggs against the turgor pressure of the pseudocoel. The cuticle and musculature of the vagina vera may enable it to act as an ovijector, expelling the eggs with some force when they are layed (Dick and Wright, 1974).

Gamete Formation

Spermatogenesis

In telogonic testes spermatogonia and spermatocytes are derived by mitosis from a single terminal germ cell or from a syncytium

Figure 4.2: The Structure of the Female Reproductive System of *Anaplectus granulosus* (A), *Meloidogyne hapla* (B), and *Hammerschmidtiella diesingi* (C). e.c., external cuticle; ov., ovary; ovd., oviduct; ut., uterus; v., vulva; v.ut., vagina uterina, v.v. vagina vera. Redrawn from Chitwood and Chitwood (1974) and taken from Wharton (1979b)

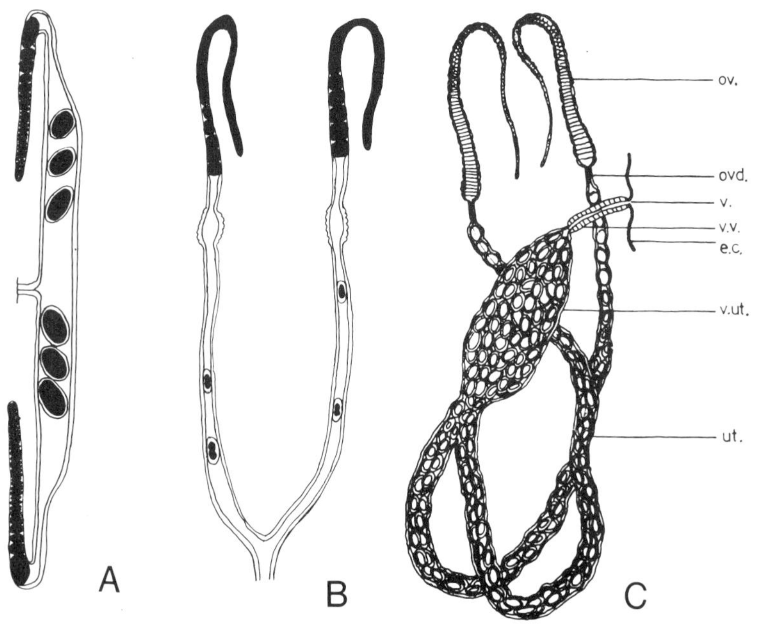

(Anya, 1976). Spermatids are formed by the two divisions of meiosis from the primary spermatocytes. These then undergo morphological changes to form mature spermatozoa. Spermatogenesis is similar in hologonic testes (Figure 4.3).

Although the structure of nematode sperm varies between different species, the sperm have features in common. Nematode sperm lack a conventional acrosome. They also have no true flagellae, and movement is by the formation of pseudopodia. The sperm contain mitochondria, microtubular structures, reserve materials (lipid droplets, refringent bodies) and complex membranous organelles (Figure 4.4).

Pseudopodial movement in the sperm of *C. elegans* is not dependent on an actin/myosin system but involves the insertion of new membrane at the tip of the pseudopodium and bulk membrane flow along the pseudopodium (Nelson, Roberts and Ward, 1982; Roberts and Ward, 1982a,b). The membranous organelles may be involved in this process.

After insemination, the sperm of *Nematospiroides dubius* change their appearance (Wright and Sommerville, 1977). Sperm isolated from the seminal vesicle are elongate but those isolated from the uterus are rounded and possess pseudopodia (Plates 4.1 and 4.2). Uterine secretions also appear to stimulate pseudopodia formation in *C. elegans* (Nelson *et al.*, 1982).

Oogenesis

Telogonic ovaries are divided into germinal, growth and maturation zones (Figure 4.5), the oogonia proliferating by mitosis in the germinal zone. There have been suggestions that oogonia are derived from a single terminal cap cell. In *Aspiculuris tetraptera*, however, the cap cell is clearly part of the ovarian epithelium (Wharton, 1979a). In most species the oogonia are attached to the rachis, a cytoplasmic extension of the cap cell. The rachis may act as a source of nutrients for the oogonia or synchronise oogonial divisions (Anya, 1976).

Laser ablation of the distal tip cell (cap cell) of the hemaphrodite gonad of *Caenorhabditis elegans* results in the arrest of mitosis in the germ cells and the initiation of meiosis (Kimble and White, 1981). The cap cell thus appears to release a diffusible substance which prevents the germ cells in its vicinity from entering into meiosis, and thus ensures the mitotic proliferation of germ cells at the distal tip of the gonad. As the germ cells pass down the gonad

Figure 4.3: Diagram of Spermatid Development (left) and the Differentiation of Spermatids into Spermatozoa (right) in *Capillaria hepatica*

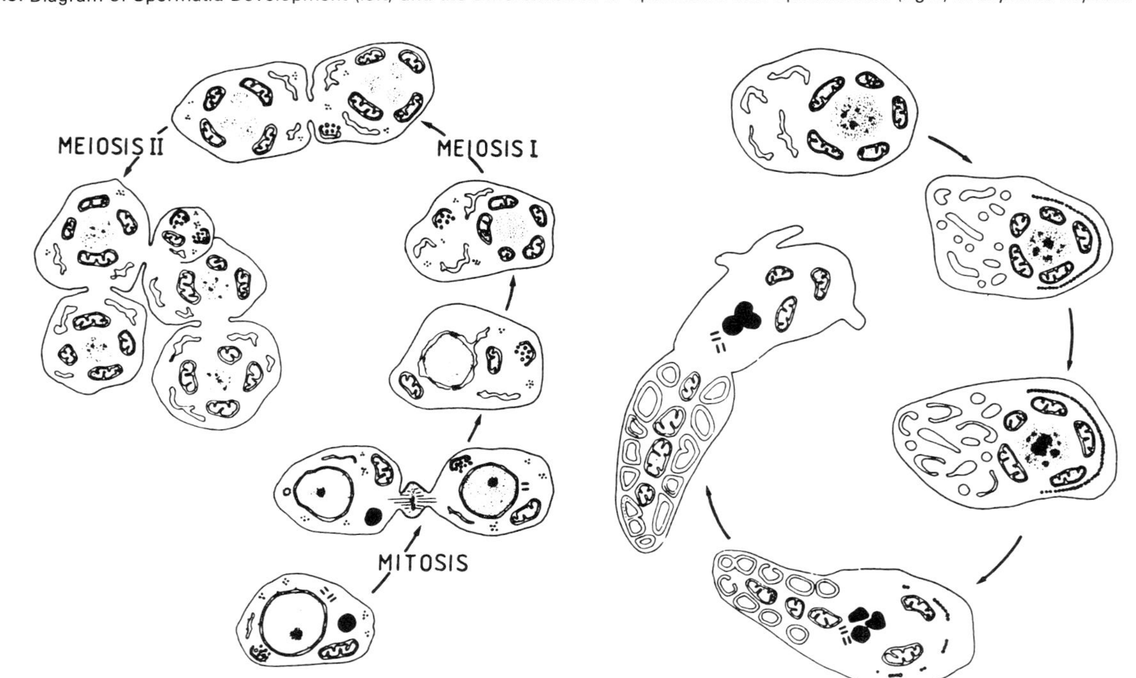

Reproduced from Neill and Wright (1973) with permission

Figure 4.4: The Structure of the Sperm of *Rhabditis strongyloides* (A), *Ascaris* sp. (B) and *Enoplus cummunis* (C). Redrawn from Chitwood and Chitwood (1974) and Maggenti (1981)

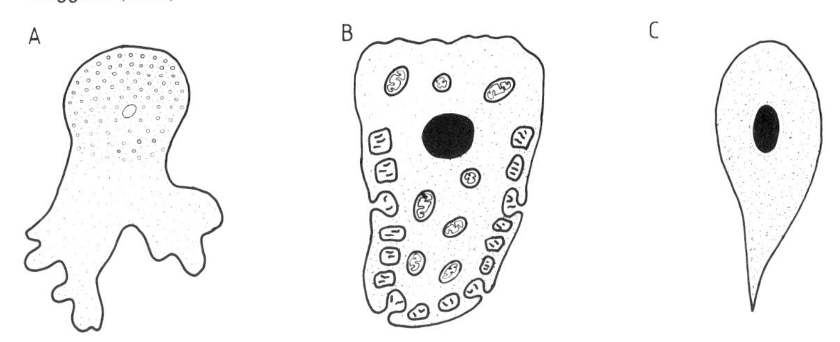

Plate 4.1: Scanning Electron Micrograph Showing Elongate Sperm Isolated from the Testis of *Nematospiroides dubius*. Arrow shows the posterior limit of cap. × 6500.

Plate 4.2: Scanning Electron Micrograph Showing Rounded Sperm Isolated from the Uterus of *Nematospiroides dubius*. × 6500

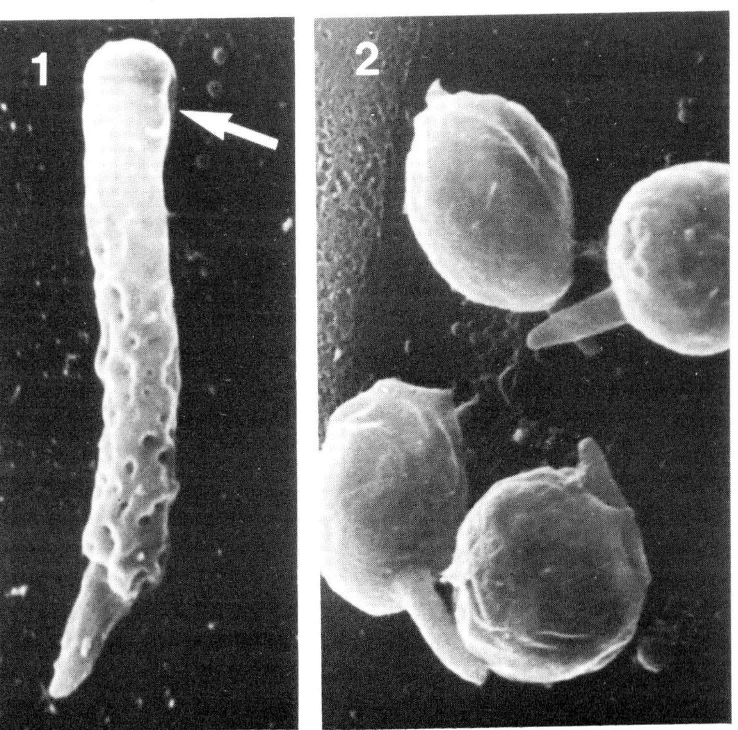

Reproduced from Wright and Sommerville (1977), with permission

Figure 4.5: The Ovary and Main Shell-forming Regions of the Female Reproductive System of *Aspiculuris tetraptera*. cc, cap cell; e, ovarian epithelium; o, oogonia; oc, oocyte; sp, sperm; us, uterine secretion

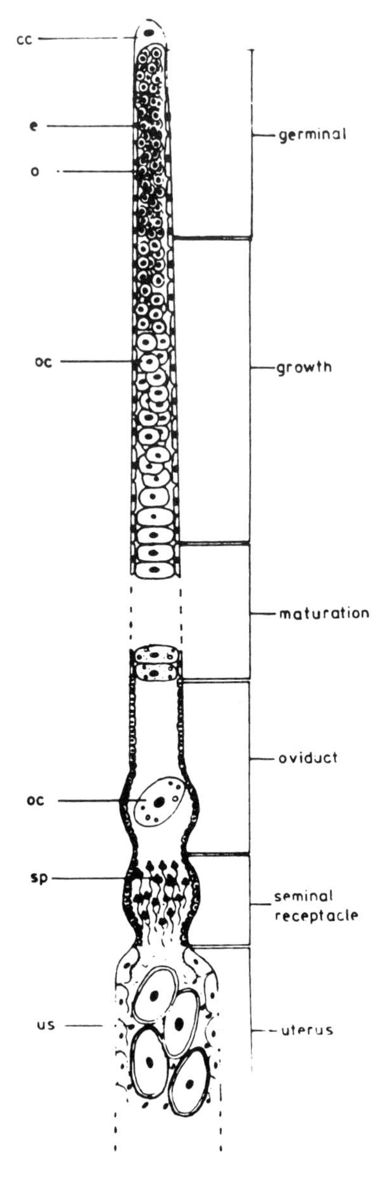

Reproduced from Wharton (1979a), with permission

they escape the inhibition imposed by the cap cell, enter into meiosis and commence their development. The oocytes undergo the first division of meiosis in the growth zone of the ovary and detach from the rachis. This division is completed upon fertilisation and the first polar body is expelled. The oocyte then undergoes the second division of meiosis and a second polar body is extruded. The egg and sperm pronuclei unite to form the zygote (Figure 4.6). Oogonial proliferation occurs throughout the length of the hologonic ovary of *Trichuris muris* but at only one point on the circumference at a time (Preston and Jenkins, 1983). The oocytes grow and mature as they pass towards the centre of the ovary (Plate 4.3).

Fertilisation also initiates egg-shell formation. The nematode egg-shell has a very restricted permeability and only allows gaseous exchange with the environment. The eggs are therefore cleidoic, and food reserves for the developing embryo must be added to the

Figure 4.6: The Main Events during Oogenesis, Spermatogenesis and Fertilisation. Oogonia and spermatogonia proliferate mitotically. The spermatids complete the two divisions of meiosis and mature into spermatozoa. The completion of the first division of meiosis of the oocyte is stimulated by fertilisation. This results in the elimination of the first polar body and initiates egg-shell formation. The egg and sperm nuclei fuse after completion of the second division of meiosis, restoring the diploid chromosome number. Original

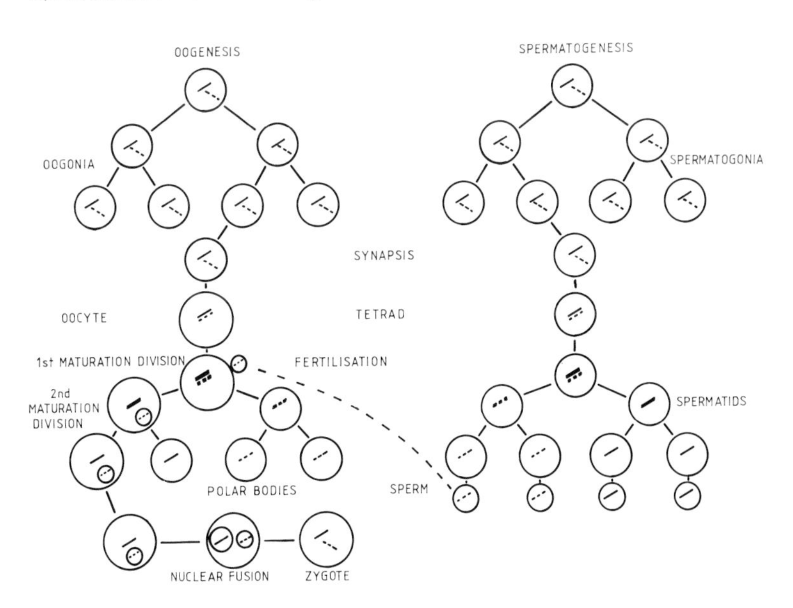

Plate 4.3: Longitudinal Section through the Hologonic Ovary of *Trichuris muris*. The oogonia (og) lie upon a basement layer (bl). The previtellogenic primary oocytes (po) are confined to one side of the ovary and contain pale-staining cytoplasm. Vitellogenic primary oocytes (vo) have a granular, darkly staining cytoplasm whereas the cytoplasm of late vitellogenic oocytes (lo) is paler staining. JB-4 section, stained with methyl green-pyronin. × 1650. Reproduced from Preston and Jenkins (1983), with permission

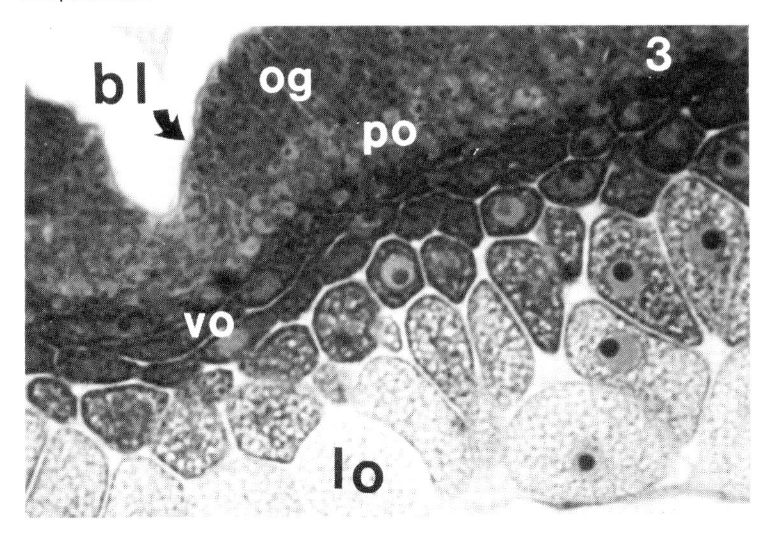

oocyte before egg-shell formation commences. The oocyte also accumulates material for the formation of the egg-shell. A variety of oocyte inclusions have been described (Table 4.1). Refringent granules, shell granules and glycogen are involved in egg-shell formation, and lipid droplets, protein bodies, hyaline granules and glycogen provide food reserves for the developing embryo.

Copulation and Fertilisation

Some plant parasitic nematodes and rather fewer animal parasitic and free-living nematodes produce substances that act as sex attractants (Green, 1980). These usually attract males to females, although female-to-male attraction has been demonstrated. The site of formation of these substances and their chemical nature are unknown. They are thought to be organic compounds and to consist of several active components. They may thus be analogous to insect pheromones.

Table 4.1: Cytoplasmic Inclusions of Nematode Oocytes

Inclusion	Chemical nature	Function
Glycogen	Glycogen	Food reserve, formation of chitinous layer
Lipid droplet	Lipid	Lipid yolk
Dense granule	Protein	Protein yolk
Hyaline granule	Protein and lipid	Lipoprotein yolk
Refringent granule	Lipid	Formation of lipid layer
Shell granule	Protein? Polyphenols?	Formation of non-chitin fraction of chitinous layer

Taken from Wharton (1979a)

The male accessory reproductive structures are involved in the sensory control of copulation, in holding the female, and in the transfer of sperm. They are very variable in structure, and are important in the taxonomy of many species. The ejaculatory duct of the male reproductive system opens into the cloaca which is situated on the ventral surface of the body and forms a joint opening with the alimentary canal. The males of most species possess one or usually two copulatory spicules. These are elongate cuticularised structures which are lodged in invaginations of the cloaca. They have a number of functions during copulation. The spicules have a complex musculature which enables them to be extended or retracted (Hope, 1974). They are used to spread open the vulva and vagina during copulation, with the aid of the other accessory reproductive structures. As the spicules do not possess a lumen through which the sperm can pass, they have not been considered to be true intromittent organs. However, in some species the two spicules can interlock to form a tube which may allow the transfer of sperm (Bird, 1976). One or more nerves run through the centre of the spicules, and some of these nerve endings are in communication with the exterior via pores in the tips. This suggests that the spicules have a sensory function before or during copulation. They are perhaps involved in mate location, detecting the vulval opening, control of entry into the vagina, and triggering the ejaculation of sperm. A cuticularised portion of the spicular pouch, the gubernaculum, may assist the extension and retraction of the spicules. In species where the spicular pouch does not lie immediately opposite the cloacal opening, a thickening of the cloacal wall, the telamon, deflects the spicules out of the cloaca. In some nematodes the male's tail is modified to form a flap-like extension called the copulatory bursa.

Copulation has rarely been observed in nematodes. In *Ditylenchus destructor* the male revolves in a spiral pattern around the female until the spicules and the vulva become aligned. The two worms are held together by the copulatory bursa which encircles the female. The spicules are inserted into the vulva and up to twenty sperm are rapidly ejaculated. The male and female then separate (Anderson and Darling, 1964). Males and females of *Nematospiroides dubius* lie with their heads in opposite directions during copulation. The male grasps the female with its copulatory bursa and inserts its spicules into the vagina. Although the males release a colourless substance from glands referred to as 'cement glands', Sommerville and Weinstein (1964) found that the worms could be easily separated, and they could not support the suggestion that they are held together during copulation by a cement. The connection between the copulatory bursa and the female worm thus appears to be purely muscular.

After copulation the sperm migrate up the female reproductive tract and are stored in the seminal receptacle, where present, or at the junction of the oviduct and the uterus. This is the site at which fertilisation occurs. In *Ascaris* the epithelial cells of this region have a distinctive ultrastructure and sperm attach to the epithelium by means of their pseudopodia (Wu and Foor, 1983). The sperm undergo changes within the uterus, including the formation of smooth endoplasmic reticulum and membrane fusion with the surface (Foor, 1974). This is related to the appearance of pseudopodial movement and is an important prerequisite for fertilisation.

Ascaris oocytes are fertilised shortly after they enter the uterus. Sperm pseudopodia make contact with the oolemma, the membranes fuse, and the sperm cytoplasm and nucleus enter the oocyte (Foor, 1968). Fertilisation stimulates the oocyte to complete its maturation divisions and initiates egg-shell formation.

The Nematode Egg-shell

Structure and Chemistry

The nematode egg-shell is a complex structure which possesses one to five layers (Wharton, 1980). In most species it consists of three layers secreted by the fertilised oocyte (Figure 4.7). The

Figure 4.7: The Variety of Structure of the Nematode Egg-shell. A. *Trichuris suis*, B. *Capillaria hepatica*, C. *Ascaris lumbricoides*, D. *Heterakis gallinarum*, E. *Porrocaecum ensicaudatum*, F. Tylenchids, G. Oxyurids. cl. chitinous layer; el, external uterine layer; il, internal uterine layer; ll, lipid layer; lpm, lipoprotein membranes; vl, vitelline layer; ul, uterine layer. Redrawn from Wharton (1980)

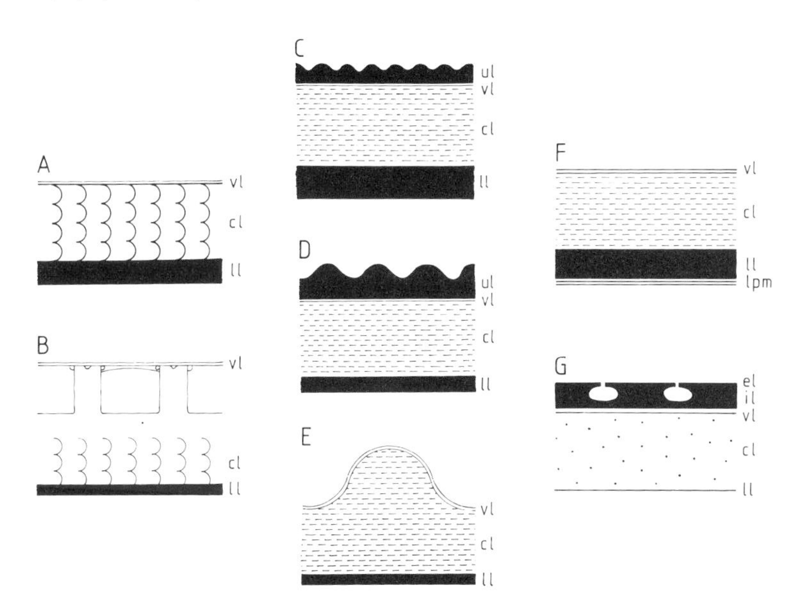

outer vitelline layer is derived from the vitelline membrane of the fertilised oocyte. The chitinous layer usually consists of a chitin/protein complex and is the only part of the nematode in which the presence of chitin has been satisfactorily demonstrated. The chitin/protein complex comprises chitin microfibrils surrounded by a protein coat to form a composite fibril (Wharton, 1980). In trichurid and capillarid egg-shells the fibrils are helicoidally arranged, giving the appearance of parabolic arcs and lamellae (Plate 4.4; Figure 4.8). In secernentian orders the fibrils are arranged in parallel or at random.

The inner lipid layer forms the main permeability barrier of the egg-shell. In *Ascaris* it contains 25 per cent protein and 75 per cent lipid. The lipid belongs to a unique class of lipids called ascarosides (Barrett, 1981). These consist of a sugar moiety (glycone), 3,6-dideoxy-L-arabinohexose and a long-chain secondary alcohol (aglycone) (Figure 4.9). Ascarosides have been found in five species of ascarid and one oxyurid, but it is likely that they

Figure 4.8: Diagram Illustrating the Bouligand Hypothesis of Helicoidal Architecture which Provides an Interpretation of the Structure Observed in the Chitinous Layer of Trichurid and Capillarid Nematodes. Fibres are parallel in each lamina but fibre direction rotates in successive laminae. Each 180° rotation of fibre direction forms one lamella (the boundary of each lamella is where the fibre direction is parallel to the surface of the section — only one lamella is shown in the diagram). The fibres of successive laminae trace out parabolic arcs when sectioned obliquely

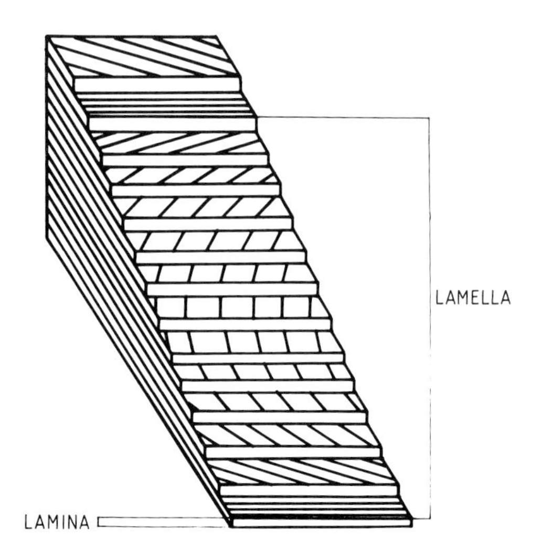

Reproduced from Wharton and Jenkins (1978), with permission

occur widely. They are responsible for the very restricted permeability of the egg-shell.

The lipid layer of tylenchid egg-shells contains lipoprotein membranes (Bird and McClure, 1976; Perry, Wharton and Clarke, 1982) which may be important in controlling the permeability of the egg-shell. They are difficult to preserve for electron microscopy and may be more widely present.

In addition to three layers produced endogenously by the oocyte, the egg-shells of some species possess one or two layers secreted exogenously by the uterus. In *Ascaris* the uterine layer consists of glycoprotein and has a mamillated appearance. The uterine layer may form filaments which attach the egg to the substrate and in aquatic species prevent the eggs settling out by entangling in vegetation (Wharton, 1983). Oxyurid egg-shells possess two uterine layers which form complex systems of spaces,

Plate 4.4: Oblique Section through the Chitinous Layer of the Egg-shell of *Trichuris suis*, Showing the Parabolic Arcs Formed by the Helicoidal Arrangement of the Composite Chitin/Protein Microfibrils. × 60 000. Original

Plate 4.5: Transverse Section through the Egg-shell of the Oxyurid Nematode, *Hammerschmidtiella diesingi*. The egg-shell consists of external uterine layer (el), internal uterine layer (il), vitelline layer (vl), chitinous layer (cl) and lipid layer (ll). The internal uterine layer contains a system of spaces which are open to the exterior of the egg-shell via pores in the external uterine layer. × 19 000

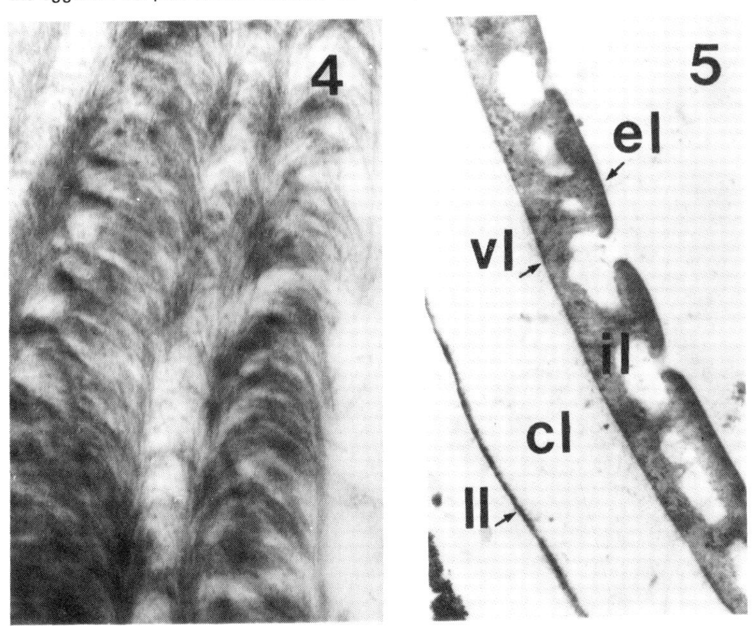

Reproduced from Wharton (1979b), with permission

open to the exterior via pores in the external uterine layer (Figure 4.10; Plates 4.5 and 4.6).

There is some evidence that the egg-shell of animal parasitic nematodes becomes tanned and that host gut enzymes are involved in this process. The mechanisms involved are unknown (Wharton, 1983).

Egg-shell Formation

After fertilisation the vitelline membrane is formed from the oolemma after the formation of a new oolemma at the surface of the oocyte cytoplasm. In *Trichuris muris* the vitelline membrane is secreted by the ovarian epithelium (Preston and Jenkins, 1983).

Plate 4.6: Freeze-etch replica of the egg-shell of *Hammerschmidtiella diesingi*. The pores in the external uterine layer are revealed. × 11 000

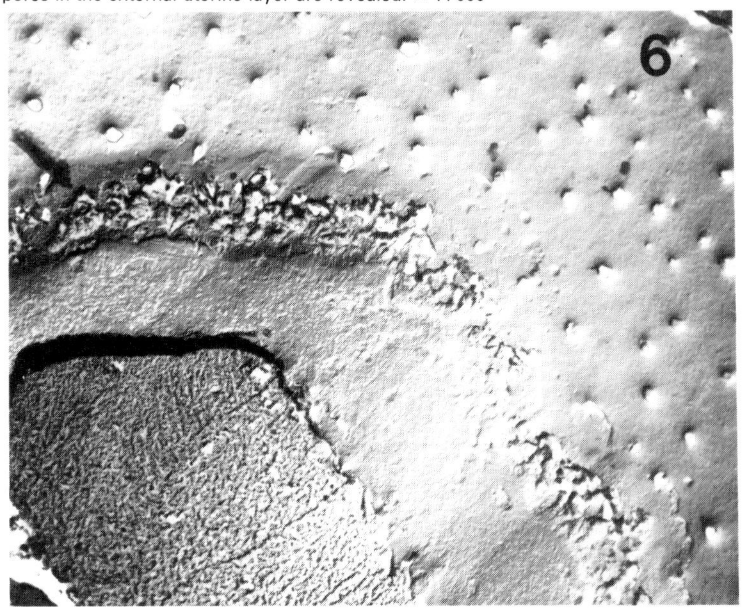

Reproduced from Wharton (1979b), with permission

Figure 4.9: Ascarosides. The lipid layer of ascarids contains a mixture of monol ascarosides (A), diol ascarosides (B), and diol diascarosides (C). The chain length of the aglycone is variable with n = an odd number between 21 and 33, depending on the type of ascaroside and the species of nematode. Redrawn from Barrett, 1981

A

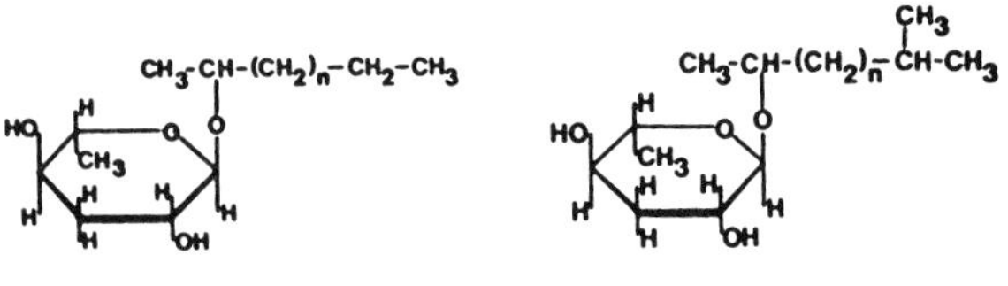

$CH_3-CH-(CH_2)_n-CH_2-CH_3$

$CH_3-CH-(CH_2)_n-CH(CH_3)-CH_3$

B

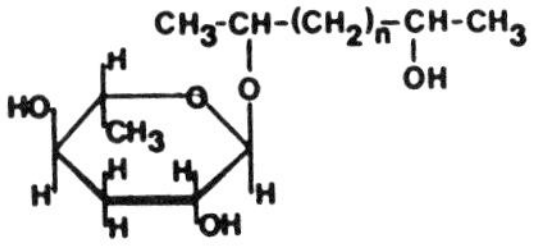

$CH_3-CH-(CH_2)_n-CH(OH)-CH_3$

C

$CH_3-CH-(CH_2)_n-CH-CH_3$

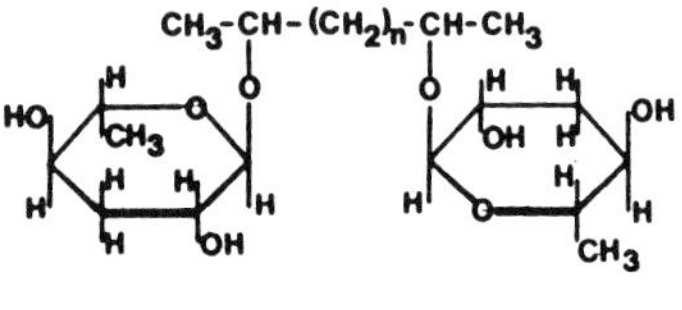

Figure 4.10: The Structure of the Oxyurid Egg-shell. The shell consists of external uterine layer (el), internal uterine layer (il), vitelline layer (vl), chitinous layer (cl), and lipid layer (ll). The uterine layers are modified to form systems of pores and spaces. A. *Aspiculuris tetraptera*, B. *Hammerschmidtiella diesingi*, C. *Syphacia obvelata* — curved side of egg, D. *S. obvelata* — flattened side of egg

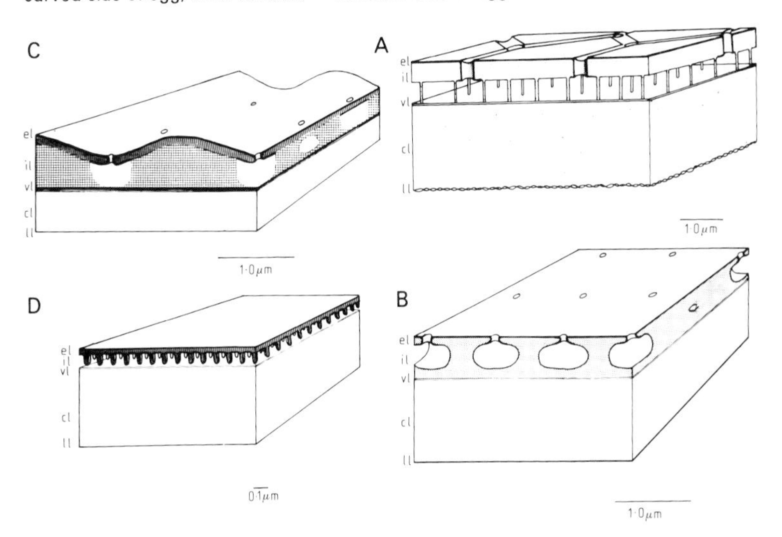

Reproduced from Wharton (1980), with permission

The vitelline membrane persists in the fully formed shell as the vitelline layer and in most species forms the outer layer of the egg-shell.

The perivitelline space between the vitelline layer and the oolemma increases in thickness as material is extruded from the egg cytoplasm to form the chitinous layer. Chitin is a polymer of N-acetyl-glucosamine and is synthesised from glycogen via glucose and glucosamine. About half the glycogen in *Parascaris* oocytes is converted into glucosamine upon fertilisation and the necessary acetate is provided by the de-esterification of ascaroside esters (Barrett, 1981). Proline is the major amino acid in the egg-shell, and proline incorporation has been demonstrated in *Meloidogyne javanica* during the formation of the chitinous layer (McClure and Bird, 1976). The chitin appears to be produced at the surface of the egg, and concentration of glycogen in the peripheral cytoplasm has been observed (Preston and Jenkins, 1983). The other components of the chitinous layer are stored as shell granules which are extruded when this layer is formed (Wharton, 1979b).

Ascarosides are stored within the oocyte as ascaroside esters. During lipid layer formation they are de-esterified to the less mobile and less permeable free ascaroside (Barrett, 1981). The material forming the lipid layer is either secreted at the surface of the egg cytoplasm or extruded from refringent granules.

The external and internal uterine layers of oxyurid egg-shells are secreted by the cells of the upper uterus. There is no evidence that the complex structure of the uterine layers is formed by a moulding process, and it may self-assemble (Wharton, 1979c). Egg-shell formation in oxyurid nematodes is summarised in Figure 4.11.

The Function of the Egg-shell

The egg-shell is one of the most resistant biological structures known and protects the embryo against environmental hazards. It is impermeable to all substances except lipid solvents, gases and perhaps liquid water (Wharton, 1980). In some animal parasites the whole of the nematode's time outside the host is spent within the protection of the egg-shell.

The chitinous layer provides mechanical strength and is resistant to chemical attack. The egg-shell of *Trichuris suis* only swells slightly after 24 h immersion in concentrated sulphuric acid (Wharton and Jenkins, 1978). The chitinous layer is a composite material, comprising chitin microfibrils surrounded by a protein matrix. A composite provides greater resistance to mechanical stress and strain than would the pure materials. The helicoidal arrangement of fibrils in trichurids and capillarids allows a higher fibre volume and makes the egg-shell resistant in all directions. The mechanical and chemical resistance of the shell may be further increased by tanning (Wharton, 1983).

The restricted permeability of the egg-shell ensures a slow rate of water loss when the egg is exposed to desiccation. This is essential for anabiotic survival (see Chapter 7) and the juveniles of several species appear to survive anabiotically within the egg (Wharton, 1980, 1982c).

Water loss can also be controlled by restricting the area through which exchange can occur. This may be the function of the pore systems found in oxyurid egg-shells. In birds, gaseous exchange across the shell is controlled by the number and physical dimensions of the pores. Similar pore systems have been found in six different phyla in the shells of eggs which may be exposed to

Figure 4.11: The Process of Egg-shell Formation in *Hammerschmidtiella diesingi*. (1) Oolemma of mature oocyte. (2) New oolemma formed upon fertilisation beneath the original, which is then referred to as the vitelline membrane. (3) The vitelline membrane forms the vitelline layer of the shell. This lifts off the surface of the egg-shell, leaving the perivitelline space between it and the oolemma. Secretion from the cells of the upper uterus forms the external uterine layer. (4) The chitinous layer forms in the perivitelline space. The shell granules are incorporated into the chitinous layer. The internal uterine layer begins to form beneath the external uterine layer from further uterine secretions. (5) The egg cytoplasm contracts away from the inner surface of the chitinous layer. Electron-dense material, produced at the surface of the egg cytoplasm, adheres to the inner surface of the chitinous layer to form the lipid layer. As secretion of the internal uterine layer continues, a system of spaces develops. This is connected to the exterior of the egg-shell via pores in the external uterine layer. (6) The formation of the lipid layer is completed. As the internal uterine layer develops, the spaces become partly filled. There is a discrete space beneath each pore. cl, chitinous layer; el, external uterine layer; il, internal uterine layer; ll, lipid layer; oo, oolemma; ps, perivitelline space; sg, shell granule; vl, vitelline layer; vm, vitelline membrane

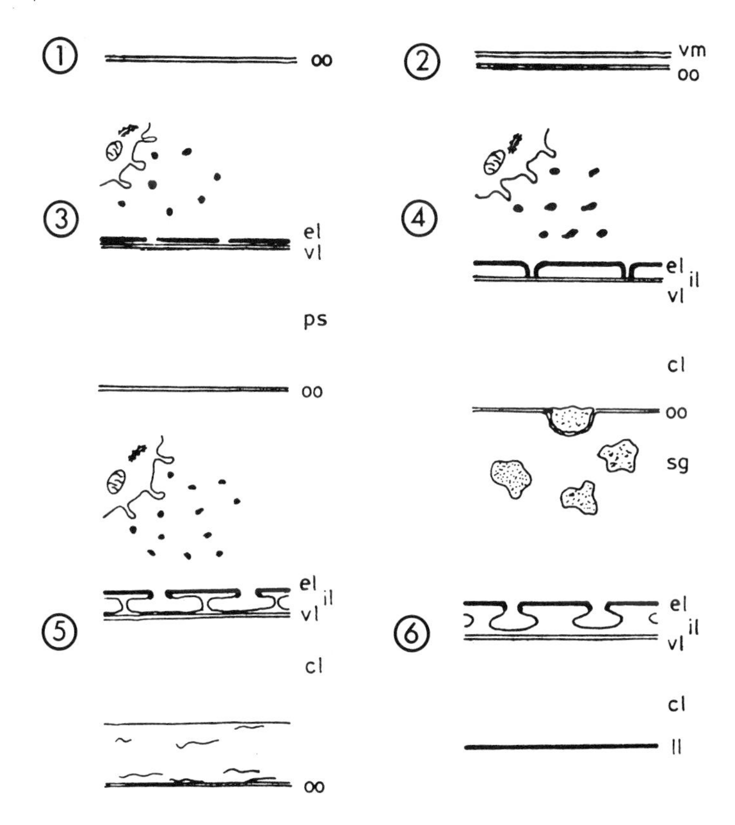

Reproduced from Wharton (1979b), with permission

desiccation. This is a striking example of convergent evolution (Wharton, 1980, 1983).

Embryonic Development

Many of the classic studies of reproduction and embryonic development in the late nineteenth century were performed on nematodes. These include the observation of zygote formation by the fusion of egg and sperm pronuclei, the halving of chromosome numbers during meiosis and their subsequent restoration upon fertilisation, the first cleavage divisions of the egg, and the separation of somatic and germ cell lines (Ehrenstein and Schierenberg, 1980). This tradition has continued in recent years with the use of the free-living nematode *Caenorhabditis elegans* as a model system to study outstanding problems in the control of development in eukaryotic organisms.

Brenner (1974) pioneered the use of *C. elegans* for genetic analysis and the study of the molecular control of development. There are now over a dozen research groups working on this nematode (Riddle, 1980). The species is easy to maintain in culture, has a constant and limited number of cells, rapid development (a generation time of 3.5 days) and reproduces hermaphroditically, facilitating the isolation and maintenance of mutant strains.

A large number of behavioural, developmental and biochemical mutants have been isolated. The spontaneous occurrence of males in *C. elegans* cultures enables gene mapping by crossing mutant hermaphrodites against wild-type males and against tester mutants (Herman and Horvitz, 1980). The ultimate aim of this work is to understand the central problem of developmental biology: how do genes function during development to produce the structures found in adult animals? Brenner (1974) has estimated that there are only 2000 somatic genes in *C. elegans* with indispensable functions. About 83 per cent of the haploid DNA content of 8×10^7 base pairs consists of unique sequences. This is the smallest DNA content reported for an animal.

The complete cell lineage of *C. elegans* has been determined (Sulston and Horvitz, 1977; Kimble and Hirsch, 1979; Sulston, Albertson and Thomson, 1980; Sulston, Schierenberg, White and Thomson, 1983). A number of conclusions emerge from this work. Cell fates are highly determined; the timing of cell divisions and cell lineages is constant from one individual to another; and the fate of any blastomere is predictable at any given moment in development. This implies that cells develop autonomously according to their own programmes, and that cell-cell interactions have a limited influence on development. Evidence for cell autonomy has also come from laser and ultraviolet ablation studies, in which specific cells are killed. Elimination of cells after the 50-cell stage has little effect on the fate of adjacent cells, suggesting that there is little cell-cell interaction (Sulston *et al.*, 1983).

A large number of cell deaths occur during development. These may reflect the formation of unnecessary cells in addition to needed cells during development. Along with other features this implies that the developmental process is inefficient, and suggests that nematodes represent an early stage in the evolution of developmental control (Sulston *et al.*, 1983).

The formation of the founder cells and the cell types derived from them is shown in Figure 4.12. Cell multiplication and the timing of the major events during the embryogenesis of *C. elegans* are shown in Figure 4.13. The pattern of development is similar in other secernentian nematodes.

Cell multiplication resumes after hatching. This mainly involves gonad development and the formation of accessory reproductive

Figure 4.12: The Formation of the Founder Cells and the Cell Types Derived from them during the Development of *Caenorhabditis elegans*. Each branch point represents a single cleavage division. Redrawn from Sulston *et al.* (1983)

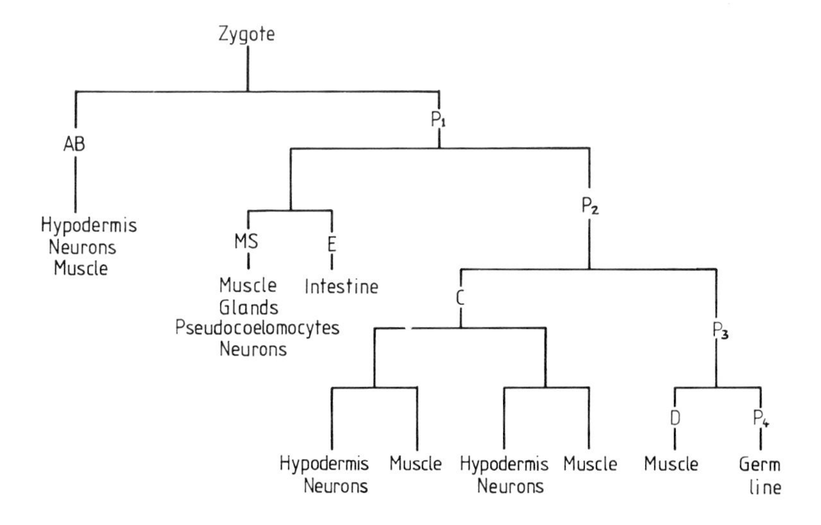

Figure 4.13: The Main Events and the Changes in the Numbers of Live Nuclei during Embryogenesis in *Caenorhabditis elegans*. Redrawn from Sulston *et al.*, (1983)

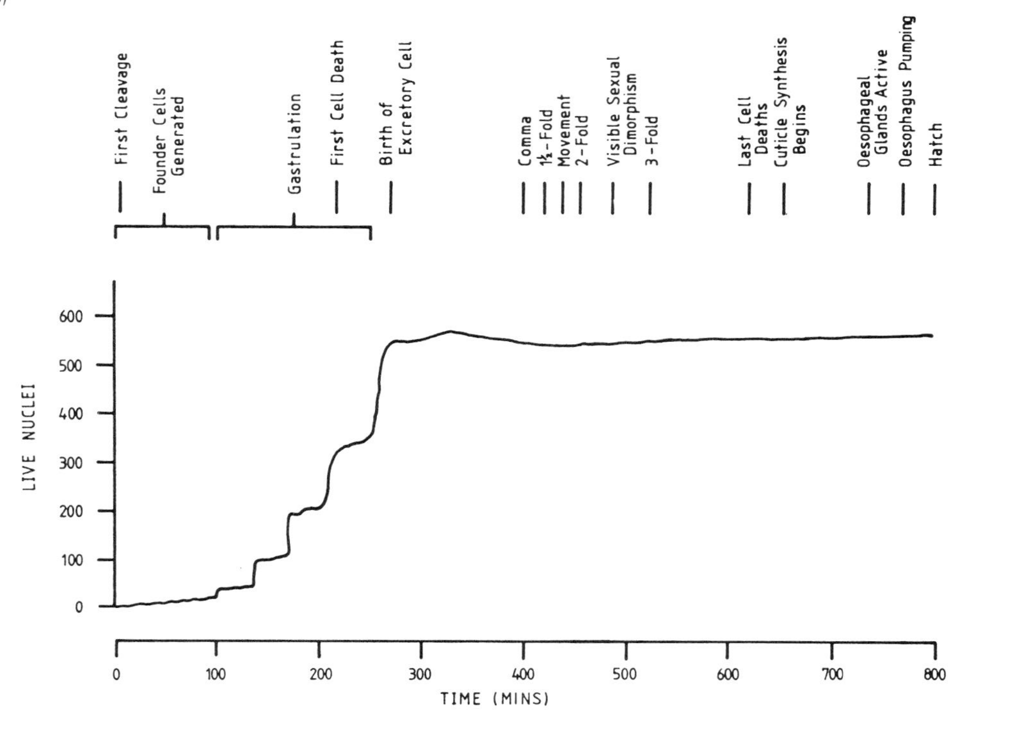

structures after the completion of the juvenile moults, but cell multiplication also occurs in the somatic musculature, hypodermis and intestine (Sulston and Horvitz, 1977). Eutely (cell numbers remaining constant once embryonic development is complete) does not, therefore, occur strictly in all nematodes but there is relatively little increase in numbers compared with other animals. Although there is some evidence of regulation in some cell lines, postembryonic development again appears to be largely autonomous (Sulston and White, 1980).

Some ascarids eliminate about 27 per cent of their DNA during the early embryonic divisions (Tobler, Smith and Ursprung, 1972). This DNA is lost from the somatic cell line whereas the germ cell line retains the full complement. This chromatin diminution has not been described in other groups of nematodes (Ehrenstein and Schierenberg, 1980). Its biological significance in ascarids is unknown but may represent the loss of redundant information from the somatic cell line.

5 PARASITISM

Despite their ecological diversity the body form of nematodes is remarkably constant, and in contrast to other groups of animals there is, at least superficially, little to distinguish parasitic from free-living species. The basic body form of nematodes is well suited to a parasitic mode of life and it has often been suggested that they are 'pre-adapted' for parasitism.

A number of features of the functional organisation of nematodes aids the colonisation of a parasitic habitat. The shape of the body is suited to longitudinal habitats such as the gut and blood vessels of vertebrates. Undulatory propulsion enables movement in a fluid medium such as gut contents or blood. A pumping oesophagus facilitates the ingestion of fluid food. A tough resistant cuticle protects the nematode against adverse features of the host environment (pH, gut enzymes, etc.) and perhaps against host reactions to infection (such as non-specific and specific immune responses).

As the basic organisation of nematodes is so well suited to parasitic habitats, it is not surprising that parasitism has arisen independently in several groups (Table 1.2). Nematodes parasitise most groups of plants, vertebrates and invertebrates, causing many diseases of economic importance.

The Evolution of Parasitism

Zooparasitic Nematodes

Parasitism of both vertebrates and invertebrates occurs in several orders of nematodes (Table 1.2). The Secernentia contains the majority of parasitic orders (Anderson, 1984). Adenophorean orders are predominantly free-living, with the parasitic species within the Enoplia presumably arising from their soil-dwelling relations.

The rarity of aquatic groups within the Secernentia also argues for the origin of parasitic species from free-living soil nematodes similar to present-day rhabditids. Some rhabditids can survive adverse conditions as a specialised 3rd-stage juvenile called a dauer larva. Transfer from a free-living to a parasitic habitat during infection of a host involves a marked change in environment, and the presence of resistant stages, such as a dauer larva, may have facilitated this transfer. The majority of parasitic species use the 3rd-stage juvenile as the infective stage.

This interpretation implies that the vertebrates were not parasitised by nematodes until after they had colonised the land (Anderson, 1984). Similarly, parasitism of invertebrate groups would have arisen among terrestrial representatives. Nematode parasites of marine invertebrates are rare. The mermithids are the only group to infect marine invertebrates, parasitising echinoderms and crustaceans. Nematode parasites that use fishes as the definitive host are a restricted fauna, representing only 11 per cent of genera parasitising vertebrates. They are similar to the parasites of terrestrial vertebrates and it seems likely that they are derived from them (Anderson, 1984).

Parasitism may have evolved from an at first casual and then progressively more intimate relationship with the host. All grades of intimacy of association are found in nematodes. Several rhabditid species use transfer hosts to transport them to fresh habitats. The dauer may penetrate the hind gut or tracheae, gaining additional protection against desiccation. It is a short step from this to penetrating the body cavity and feeding parasitically, as occurs in *Parasitorhabditis* spp. (Osche, 1963).

Phytoparasitic Nematodes

Plant parasitism has arisen independently in three orders of nematodes: in the Dorylaimida of the class Adenophorea and in the Tylenchida and Aphelenchida of the class Secernentia (Siddiqui, 1983; Table 1.2). The presence of a stylet is a feature common to all plant parasitic nematodes but is also found in some insect parasites and predacious species. It may have evolved to reach inaccessible micro-organisms and later in predacious species and fungal- and algal-feeders. Some free-living dorylaimids are often found with bacteria in the stylet lumen (Siddiqui, 1983). The stylet is thus a pre-adaptive feature, although it is an important adaptation for plant parasitism.

Plant parasites are thought to have arisen from terrestrial soil-dwelling nematodes. There are a few marine plant parasites but these are probably derived from terrestrial forms (Siddiqui, 1983). Various routes for the development of plant parasitism have been proposed (Maggenti, 1971; Siddiqui, 1983). These include evolution from nematodes feeding on ectoparasitic fungi, those feeding on endoparasitic fungi, and fungal-feeders that utilised an insect transport host.

The intimacy of the relationship between nematodes and plants varies. Some browse on the root surface, feeding on bacteria and epidermal cells. Sedentary endoparasites have complex morphological and physiological interactions with their hosts and some consider that only nematodes in this latter group are true parasites (Siddiqui, 1983).

Host/Parasite Relationships

Plant Parasitic Nematodes

Ectoparasitic nematodes feed by penetrating plant cells with their stylets and ingesting the contents. Feeding may be confined to the epidermal cells with the nematode moving from cell to cell. Such activity may cause only superficial damage to the plant, although it may lay the plant open to secondary bacterial and fungal infections. *Pratylenchus* spp. can feed for several days on a single cell without any apparent destructive effect (Wyss, 1981).

The initial phase of feeding involves penetration of the cell by the stylet, followed by a period of inactivity (Plate 5.1). The nematode is thought to be preparing for salivation during this period and, in some cases, fluid has been observed moving towards the stylet before it is pumped into the cell. Salivary secretions may partly digest the cell contents before ingestion. The cell contents are drawn up the stylet by the pumping of the oesophagus, and the stylet is then withdrawn.

Some species possess longer stylets and can feed on plant tissues deeper than the epidermis. These have a more complex relationship with the plant, inducing gall formation via increases in the number and size of cells as well as causing necrosis (cell death). It is not known how the nematode induces these changes in the plant, although it is assumed that substances within the salivary secretions act as a chemical trigger (Wyss, 1981).

Although ectoparasitic species cause little damage to the host themselves, they may act as vectors of important plant diseases. Nematodes are known to act as vectors of some 20 plant viruses, which cause diseases such as ringspot of tobacco, tomato, rasp-

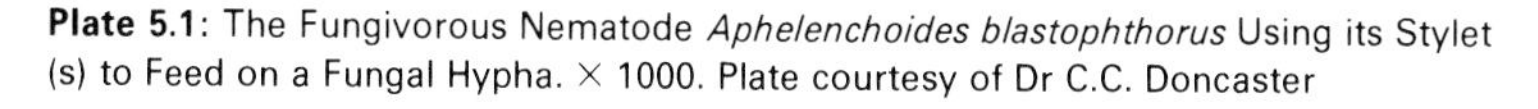

Plate 5.1: The Fungivorous Nematode *Aphelenchoides blastophthorus* Using its Stylet (s) to Feed on a Fungal Hypha. × 1000. Plate courtesy of Dr C.C. Doncaster

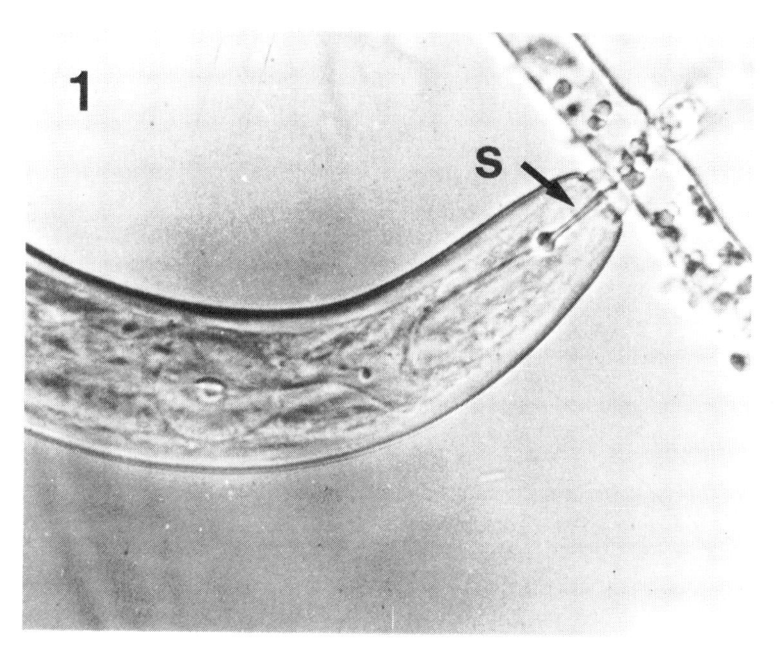

berry and strawberry, tobacco rattle, and pea early browning (Taylor and Brown, 1981). Various *Xiphenema, Longidorus* and *Trichodorus* spp. act as vectors, acquiring the virus from infected plants during feeding. The virus particles are absorbed on to the lumen of the stylet during feeding, and detach in response to the secretions produced during salivation. There is a high degree of specificity in virus transmission. *Xiphenema diversicaudatum* will transmit arabis mosaic virus but not raspberry ringspot virus, even if both are present.

Endoparasites may also act as vectors of plant diseases. *Anguina* spp. transmit *Corynebacterium*, causing annual ryegrass toxicity (Bird, 1981), and rickettsia-like micro-organisms are associated with *Globodera rostochiensis* (Walsh, Shepherd and Lee, 1983).

Endoparasitic nematodes may be migratory or sedentary. Migratory root nematodes enter the root tissue and cause damage both by their feeding activities and by the release of toxins. Other migratory endoparasites infect the aerial parts of plants, including the stems, leaves, buds and reproductive tissues. Some of these have eliminated soil stages altogether and are transmitted by insects.

Sedentary endoparasitic nematodes form complex relationships with their hosts (Table 5.1). The nematode induces the formation of a syncytium (e.g. *Globodera* spp.) or of a giant cell (e.g. *Meloidogyne* spp.) in which it inserts its stylet and feeds. In *Tylenchus semipenetrans* six to ten modified cells (nurse cells) surround the head of the nematode, from which it feeds in turn.

Giant cells are formed by repeated nuclear mitoses in the absence of cell division, resulting in a large multinucleated cell (Figure 5.1). Giant cell enlargement continues for 2 to 3 weeks so that the cell eventually contains 32 nuclei or more (Jones, 1981). In *Rotylenchus macrodoratus* a giant cell develops with a single, very large nucleus. The giant cell may possess extensive areas of wall ingrowths, particularly in areas adjacent to vascular tissues such as xylem and sieve elements (Plate 5.2). Syncytia are formed by the walls between adjacent cells breaking down and the protoplasts fusing (Plate 5.3). Wall ingrowths are again formed where the syncytium contacts vascular tissue.

Giant cells and syncytia divert plant nutrients to the feeding site of the nematode. Cell wall ingrowths similar to those found in nematode-modified cells are present in normal plant cells which act as 'transfer cells'. These are found in a variety of locations and may be absorptive or secretory. The wall ingrowths are thought to enhance solute uptake by increasing the area of cell membrane and by establishing a hydrogen-ion and solute gradient within the ingrowth. This acts as a hydrogen-ion co-transport system driven by the potential difference of H^+ ions inside and outside the cell (Figure 5.2).

There is good evidence that the nematode is responsible for producing these modifications in host cells. Different nematode genera infecting the same host produce their typical modifications (giant cell, syncytium, nurse cells). The stimulus thus comes from the nematode and is not just a response by the plant to damage. The nature of the substance secreted by the nematode which causes these changes is unknown. Auxins, other plant hormones, histones and enzymes have all been suggested. Interestingly, the effect of caffeine on plant cells mimics giant cell formation by preventing cell division. Caffeine inhibits cAMP phosphodiesterase, resulting in increased cAMP levels. It can also sequester Ca^{2+} ions.

Table 5.1: A Comparison of the Changes Induced in Plant Host Cells by Endoparasitic Nematodes

	Structure	Mitotic stimulation	Nucleus enlargement	Multinucleate cytoplasm	Cell expansion	Wall ingrowths	Plasmodesmatal frequency
Heterodera spp.	S	–	+	Yes	+	+	Low
Globodera spp.	S	–	+	Yes	+	+	Low
Meloidogyne spp.	GC	+	+	Yes	+	+	Low
Meloidodera floridensis	GC	+	+	?	+	?	?
Rotylenchus reniformis	S	–	+	Yes	+	–	High
Rotylenchus macrodoratus	GC	–	+	No	+	+	?
Naccobus spp.	S	+	+	Yes	+	–	High
Tylenchulus semipenetrans	NS	–	+	No	–	?	?

S = Syncytium, GC = giant cell, NS = nurse cells, + = present, – = absent
Taken from Jones (1981)

Figure 5.1: Giant Cell Formation in *Meloidogyne* sp. Nuclear division occurs without cell division. This results in multinucleate cells undergoing synchronous mitoses. Vascular elements differentiate adjacent to the giant cells. Redrawn from Jones (1981)

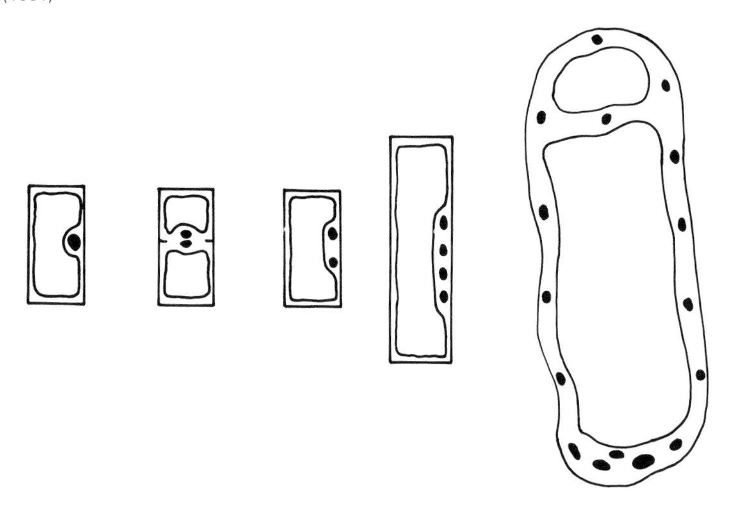

Alternatively, giant cell and syncytium formation may not be induced by the nematode, and the plant may simply be responding to the nutrient sink caused by the nematode (Jones, 1981).

Animal Parasitic Nematodes

Nematodes parasitise many invertebrates and most vertebrates. They cause a number of economically important diseases of humans and their domestic stock (Table 5.2; see also Table 1.2). The alimentary canal is the principal site for nematode parasites, although they also infect almost every other part of the body.

Parasites of the Alimentary Canal. The gut possesses many advantages as a parasitic site but it also presents a number of hazards. The parasite gains a constant and readily assimilable source of nutrients and protection against the hazards of the external environment. However, the gut also undergoes constant movement, secretes a battery of enzymes which could attack the parasite, has a pH varying from 1.5 to 8.4, has fluctuating redox potentials, is at least partially anaerobic and may develop an immune reaction against the parasite. The nematode must prevent damage by host enzymes and varying physicochemical conditions,

Plate 5.2: Scanning Electron Micrograph of the Junction between a 15-day-old Giant Cell Induced by *Meloidogyne incognita* on Balsam, and Xylem Elements (x). The internal wall of the giant cell is covered by wall ingrowths (i). × 1250. Reproduced from Jones and Dropkin (1976) with permission

Plate 5.3: Scanning Electron Micrograph of Syncytium Induced by *Heterodera glycines* on Soybean Root Cortex. The syncytium is formed by the development of cell wall perforations (p) in the root tissue. × 700

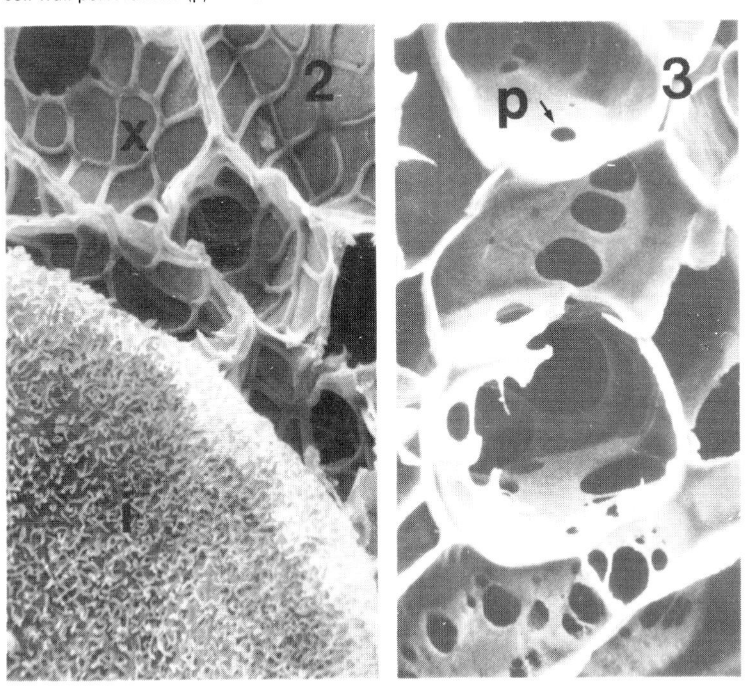

Reproduced from Jones and Dropkin (1975) with permission

maintain its position within the gut, cope with low oxygen tensions and mitigate the host's immune reaction.

Although the mechanisms involved are unclear, nematodes appear to protect themselves against host enzymes; dead nematodes are rapidly digested. The resistant and impermeable cuticle may provide protection against chemical attack. Antienzymes which inhibit the host's digestive enzymes have been found in several species but there is no evidence that they are released into the gut (Barrett, 1981).

It is not easy to determine how nematodes maintain their position within the gut as they are difficult to observe *in situ.*

Figure 5.2: Postulated Transfer Mechanism of the Cell Wall Ingrowths of Nematode-induced Giant Cells and Syncytia. A H^+ gradient develops from the tip to the base of the ingrowth and a solute gradient develops in the opposite direction. Transfer is driven by a H^+ ion co-transport mechanism. The driving force is thus greatest where the solute concentration is lowest and vice versa. This ensures efficient uptake throughout the length of the ingrowth. Redrawn from Jones (1981)

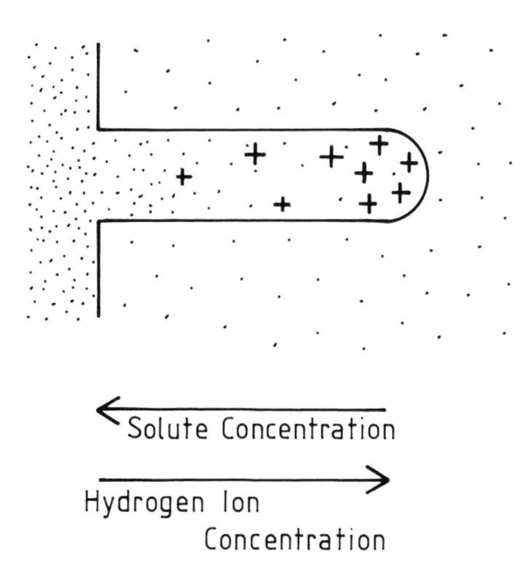

Recent attempts include the observation of *Nippostrongylus brasiliensis* through the walls of surgically exposed loops of the intestine of anaesthetised rats (Croll and Smith, 1977) and the use of an endoscope (Nicholls, Lee and Sharpe, 1985). It is thought that gut nematodes actively maintain their position by undulatory propulsion. *Nippostrongylus* may use a single spiral wave to brace itself against the villi (Lee and Atkinson, 1976). Some blood-feeders attach to the villi using their buccal capsule, and some species secrete an acetylcholinesterase which acts as a 'biochemical holdfast' by damping down local peristaltic movements (Lee and Atkinson, 1976).

The oxygen tension within different vertebrate tissues varies considerably (Table 5.3). Oxygen tensions within the gut are low, and although intestinal nematodes may have some oxygen available to them, they must survive periods of anaerobiosis. Blood-feeders may be able to extract some oxygen from the blood passing through the intestine of the nematode or they may live close to the villi where oxygen tensions are higher than those in the lumen. Some species possess haemoglobin in their cuticle and in other sites, which aids the efficiency of oxygen uptake (Atkinson, 1976).

Table 5.2: Estimates of the Numbers (in millions) of Nematode Infections of Humans

Nematode	Tropics and subtropics				Temperate zones			
	Africa	Asia (−USSR)	Central and South America	Oceania	North America	Europe (−USSR)	USSR	World totals
Ascaris lumbricoides	159	931	104	1	5	39	30	1269
Hookworms	132	685	104	2	3	2	4	932
Enterobius vermicularis	24	136	40	1	29	75	48	353
Trichuris trichuria	76	433	94	1	1	41	41	353
Trichinella spiralis	1	—	3	—	35	5	2	46
Other intestinal nematodes	9	49	21	<1	1	1	3	84
Wuchereria bancrofti and *Brugia malayi*	59	300	22	2	—	—	—	383
Other filariae	178	57	39	—	—	—	—	274
Total infections	638	2591	427	7	74	163	128	4028

Taken from Peters (1978)

Table 5.3: Oxygen Tensions in Vertebrate Tissues

Habitat	Species	Oxygen tension (mmHg)
Swim bladder	Fish	>760
Skin	Mammals	50-100
Subcutaneous tissues	Mammals	20-43
Arterial blood	Mammals, birds and fish	70-100
Venous blood	Mammals and birds	40-66
	Fish	15-20
Peritoneal cavity	Mammals	28-40
Pleural cavity	Mammals	12-39
Urine	Mammals	14-60
Bile	Mammals	0-30
Stomach	Mammals (gases)	0-70*
Small intestine	Mammals (gases)	0-65*
	Mammals and birds (contents)	0.5-30†
Large intestine	Mammals (gases)	0-5

* High values due to swallowed air; † highest values near mucosa
Taken from Barrett (1981)

Although aerobic pathways persist in parasitic nematodes, the main pathways of carbohydrate breakdown used are anaerobic and result in the formation of organic end-products (Barrett, 1981). Carbohydrate catabolism has been most extensively studied in *A. lumbricoides* (Figure 5.3). Glycogen is broken down to phosphoenolpyruvate by glycolysis. Carbon dioxide fixation is catalysed by the enzyme phosphoenolpyruvate carboxykinase to give oxaloacetate, which is reduced to malate. The malate then enters the mitochondria where it undergoes a dismutation reaction. Part is reduced to succinate and the rest is oxidised to pyruvate. Pyruvate and succinate are metabolised further to form a complex mixture of organic acids which are excreted. These pathways can operate completely anaerobically, although similar end-products are produced under aerobic conditions.

The relative efficiencies of these anaerobic pathways are low compared with those for the complete aerobic breakdown of glucose via glycolysis, the tricarboxylic acid cycle and the cytochrome chain. In the nutrient-rich environment provided by the gut, parasitic helminths may be under less pressure to maximise the efficiency of energy utilisation than their free-living relations (Barrett, 1984).

The presence of a resistant cuticle with restricted permeability means that nutrient uptake must occur via the gut. In contrast to

Figure 5.3: Carbohydrate Catabolism in *Ascaris lumbricoides* Muscle. Redrawn from Barrett (1981)

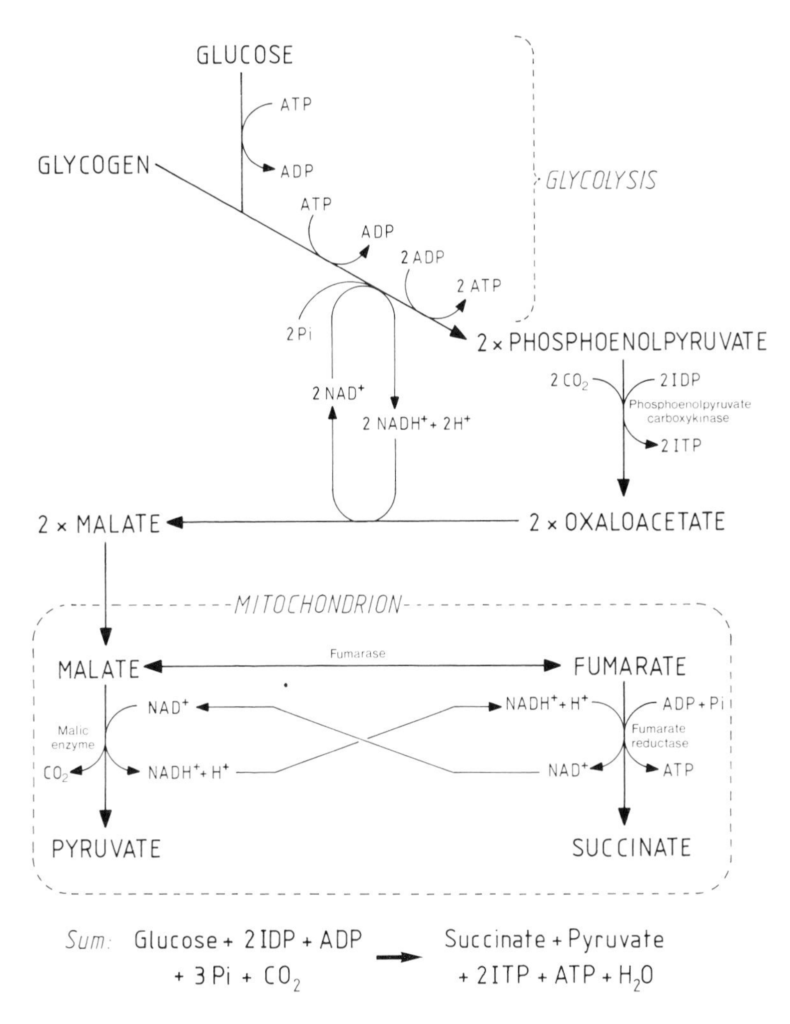

cestodes, digeneans and acanthocephalans, there is little evidence for transcuticular nutrient uptake except in nematodes inhabiting the blood of vertebrates or the haemocoel of insects.

Nematodes exhibit a variety of feeding strategies within the vertebrate gut. Many feed directly on the gut contents, others by abrading the mucosa and ingesting epidermal cells and blood. The hookworms, *Ancylostoma* and *Necator*, draw a plug of intestinal mucosa into their buccal capsule. The tissue is broken down and blood is rapidly pumped through the intestine of the nematode so that most of it goes undigested. Blood-feeding nematodes can cause serious anaemia in their hosts. Intestinal capillarid and trichurid nematodes partially bury their anterior ends in a tunnel-like construction in the gut wall (Figure 5.4). A series of cells adjacent to the oesophagus called stichocytes are thought to secrete enzymes. These digest the mucosa which is then ingested via the mouth (Jenkins, 1970).

The juvenile stages of some intestinal nematodes undergo extensive migrations throughout the body before reaching their final site. The juveniles of *A. lumbricoides* penetrate the gut wall after hatching and migrate via the blood vessels to the liver, heart and lungs. From here they pass up the trachea and back into the alimentary canal. Why juveniles that hatch in the gut and return to the gut to complete their development to adults undergo these complex migrations is unknown. Perhaps they require substances for development that cannot be obtained within the gut. Nematode parasites that infect via the oral route are thought to have evolved from those with skin-penetrating juveniles. The migrations may, therefore, be a remnant of those necessary for a skin-penetrator to reach its final site in the gut. Wilson (1982) has pointed out, however, that descriptions of juvenile migrations of skin-penetrating nematodes are based on early work on *Nippostrongylus*

Figure 5.4: Attachment of *Trichuris suis* to the Caecal Mucosa of the Pig. The thin oesophageal region lies within a tunnel-like construction in the mucosa. The mouth and the stouter posterior region lie unattached. at, attachment tunnel; hm, host mucosa; lp, lamina propria; oe, oral end; or, oesophageal region

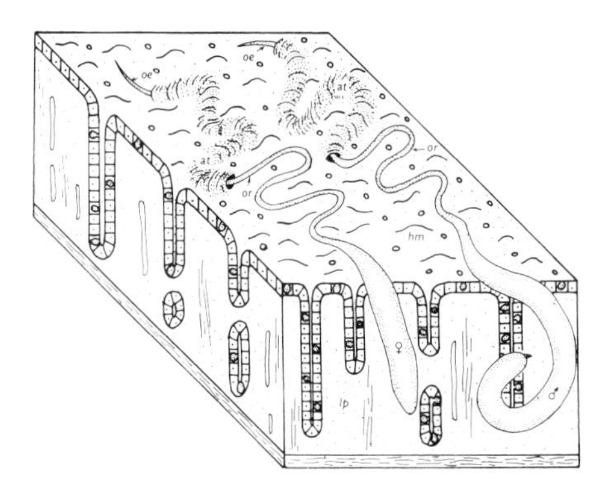

Reproduced from Jenkins (1970) with permission

brasiliensis. There is little evidence of migration via the lungs in other species.

Parasites of the Blood and Lymphatic Systems. The blood is potentially an attractive parasitic habitat. It provides a rich source of nutrients and of oxygen, both dissolved in the plasma and in blood cells. Filarid nematodes have successfully colonised the blood, both as 1st-stage juveniles (microfilariae) and as adults in associated sites in the lymphatic system and subcutaneous tissues. Filarids cause important parasitic diseases of humans (Table 5.4). The World Health Organization estimates that some 600 million people suffer from filarid infections and has designated them a priority area for research.

Microfilariae lack a functional gut and appear to take up nutrients via the cuticle. Adults possess a mouth and an apparently functional gut, and take up high-molecular-weight compounds via this route *in vitro*. Low-molecular-weight compounds can, however, be absorbed through the cuticle. Transcuticular uptake of glycine, adenosine, D-glucose and L-amino acids has been demonstrated *in vitro* in *Brugia pahangi* (Howells, 1980). The epicuticle and the outer hypodermal membrane beneath the cuticle of *Onchocerca gutturosa* are highly folded. These both have the appearance of highly absorptive surfaces.

Other Parasitic Sites. In addition to the alimentary canal and the vascular systems, nematodes infect many other sites within vertebrates (Kennedy, 1976). Parasites of the respiratory system may be found within the lung parenchyma and alveoli, the bronchi and bronchioles or the trachea. *Syngamus trachea* inhabits the trachea of various birds, including the chicken. The lungworms of sheep and cattle, *Dictyocaulus viviparus* and *D. filaria*, are found in the bronchioles and bronchi of the host. The presence of the nematode

Table 5.4: The Major Filarid Nematodes Causing Diseases of Humans

Filarid	Site (adults)	Disease
Wuchereria bancrofti	Lymphatic system	Elephantiasis
Onchocerca volvulus	Subcutaneous tissue	River blindness
Brugia malayi	Lymphatic system	Malayan filariasis
Loa loa	Subcutaneous tissue	Eye worm

within the lungs results in a host reaction which causes the lungs to fill with fluid, increases the breathing rate and stimulates coughing (parasitic pneumonia or 'husk'). This often results in the death of the host. Control of *Dictyocaulus* provides one of the few examples of successful vaccination against a helminth parasite. Injection of irradiated juveniles will evoke a strong protective immune response. There is little information on the energy metabolism of lung parasites. They have considerably more oxygen available to them than do those in other parasitic sites, and might therefore be expected to support an aerobic metabolism.

The musculature is an unusual parasitic site but is used by *Trichinella spiralis* for the encystment of the infective juvenile. The infection is transmitted to a new host when the infected muscles are eaten. Nematodes may appear in the eye and brain as the result of juvenile tissue migrations, and do not utilise these sites as permanent niches. The juveniles are tolerated when they are alive, but upon death they evoke an intense inflammatory reaction which, if it occurs in the retina or optic nerve, can cause blindness.

The peritoneal cavity, the liver, the kidney and the subcutaneous tissues are also used as parasitic sites by nematodes. The functional organisation of nematodes has enabled them to adapt to a very wide range of parasitic habitats.

Parasites of Invertebrates

Nematodes parasitise most invertebrates, although infections of marine invertebrates are relatively rare (Anderson, 1984). Those infecting insects have received the most attention, because of their potential for biological control (Poinar, 1979). Members of the genus *Neoaplectana* carry symbiotic bacteria. The juveniles can enter the insect via various body openings, including the mouth, anus and spiracles, and penetrate the body cavity via the gut or tracheae. The bacteria are released and multiply rapidly within the haemocoel, causing a massive septicaemia and the death of the host. The bacterium and the nematode are dependent upon one another for normal development, but it is not known whether the nematode feeds directly on the bacteria or on insect tissue broken down by bacterial enzymes, although bacteria are found in the intestine. *Neoaplectana* spp. have been used in field trials against a number of insect pests and are being marketed as a biological insecticide.

The adults and 4th-stage juveniles of mermithid nematodes are free-living within the soil. The free-living stages are non-feeding, relying on reserves accumulated by the parasitic 2nd- and 3rd-stage juveniles. The parasitic stages infect the body cavity of molluscs, echinoderms, insects and other arthropods. The intestine enlarges and forms a solid mass of cells which acts as a food storage organ (the trophosome). The 3rd-stage juvenile emerges though the body wall, killing the host. Mermithids infect a number of insect pests, and their potential as a biological control agent has been investigated (Poinar, 1979).

The arthropod haemocoel provides a rich supply of nutrients, and some nematodes can absorb these directly. *Bradynema* sp. lacks a cuticle and absorbs nutrients through the microvillous hypodermis (Riding, 1970). In *Sphaerularia bombi* the entire uterus of the female is everted through the vagina. It expands enormously and dwarfs the body of the female. Nutrient uptake occurs via the surface of the uterine cells (Poinar, 1983).

Nematodes also infect the reproductive system, tracheae and gut of insects. Oxyurid nematodes infect the gut of a variety of invertebrates, as well as that of vertebrates. The life cycle is direct, the host being infected by ingesting an egg containing an infective juvenile.

Host Reactions to Nematodes

Vertebrate Defence Reactions

Nematodes parasitic in the gut, blood system and other sites are foreign to the animal and are recognised as such by the host's defence system. The host responds specifically to parasite antigens via the immune system. In addition, the parasite may induce a number of non-specific responses which act to its detriment.

Physicochemical conditions within the gut can be altered by the presence of intestinal nematodes. *Haemonchus contortus* infections cause changes in pH and Na^+ levels within the abomasum of sheep (Mapes and Coop, 1973). Changes in these conditions could act to the detriment of the nematode and to that of concurrent infections with other parasites. Nematode parasites are often dependent on specific cues (CO_2, pH, temperature changes, etc.) from the host for development to proceed. These cues may be disrupted by existing infections and in hosts that acquire low levels of infection. Inflammation can result from physical irritation caused by the parasite, which will again result in changes in the environment of the nematode and result in its elimination, particularly if the response stimulates the secretion of intestinal mucus (Barriga, 1981).

The specific immune reactions of vertebrates to parasite antigens include cell-mediated responses and the production of antibodies (humoral immunity). Antibodies belong to the immunoglobulin (Ig) fraction of serum proteins and are divided into five classes: IgA, IgD, IgE, IgG and IgM. The lymphocytes involved in both cell-mediated and humoral immunity are derived from stem cells which, in higher vertebrates, are produced by the bone marrow (Figure 5.5). Lymphocytes involved in humoral responses develop from stem cells which are processed by the bursa of Fabricius in birds or its unknown equivalent in mammals,

Figure 5.5: Diagram Showing the Two Arms of the Vertebrate Immune System. Original

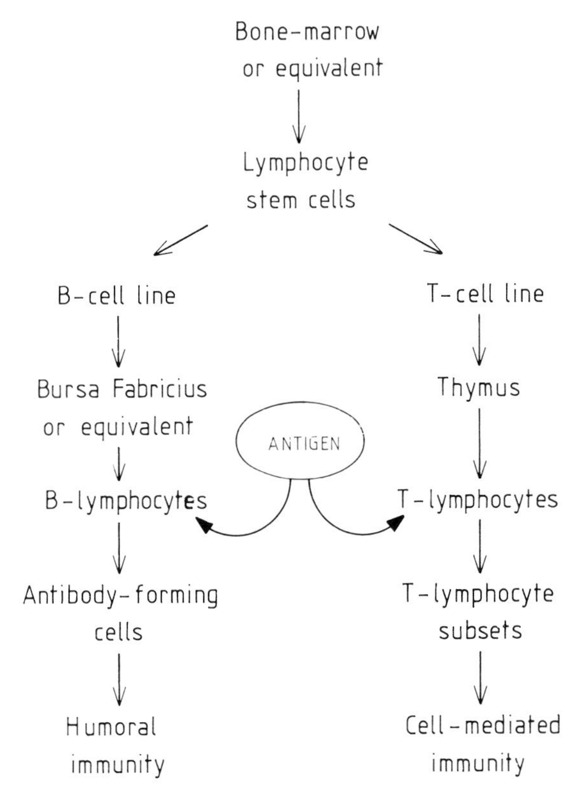

giving rise to B-lymphocytes. These respond to antigens, which need to be presented by macrophages and helper T-cells, by undergoing rapid division and then differentiating into plasma cells which secrete antibody into the blood and surrounding tissue. Antibodies bind to their target antigens, activating complement (cytotoxic serum proteins) which damages the surface of the invading organism or acts as a recognition site for defensive cells (macrophages, eosinophils, neutrophils).

In cell-mediated immunity, stem cells are processed in the thymus gland to produce T-lymphocytes. T-cells may respond to antigen in a variety of ways. They may act as effector cells themselves (cytotoxic T-cells) or indirectly by producing lymphokines, which are chemical signals attracting a variety of cells (e.g. macrophages, eosinophils), which are activated close to the antigen source (delayed hypersensitivity-type T-cells).

T-cells can also act as helper or suppressor cells mediating the B-cell or T-cell response. Some are retained in a resting phase and act as memory cells. T- and B-cell responses may be dependent upon the processing and presentation of antigens by macrophages and other cells.

Antigenic stimulation is provided by the surface of the nematode cuticle (surface antigens) and by the release of excretory and secretory products (excretory/secretory antigens). The host responds to nematode antigens by antibody production, even in the case of intestinal infections. It is known both that antigens can be taken up across the gut wall and that antibodies are secreted into the gut lumen (Wakelin, 1984a). Antigen transport across the gut wall is facilitated by specialised regions containing aggregations of lymphoid tissue (gut-associated lymphoid tissue — GALT). These include Peyer's patches and the appendix (Befus and Bienenstock, 1984). To what extent antigen uptake is confined to these regions is unknown; some macromolecular uptake occurs across the general intestinal mucosa. B-cell responses to intestinal infections include all the major classes of antibodies but only IgA and IgM are actively secreted into the lumen of the gut and are resistant to degradation by gut enzymes.

Most evidence for protective immunity against intestinal nematodes has come from laboratory infections (Table 5.5). The best known example in a natural infection is the 'self-cure' phenomenon observed in sheep infected with *H. contortus.* The existing adult parasite population is rapidly expelled after the

Table 5.5: Spontaneous Cure in Laboratory Infections with Nematode Parasites

Nematode	Host	Thymus dependency	Transfer of response		Associated with	
			Serum	Cells	Reaginic antibody	Inflammation
Nippostrongylus brasiliensis	Rat/mouse	+	+	+	+	+
Trichinella spiralis	Rat/mouse	+	+	+	+	+
Trichuris muris	Mouse	+	+	+	–	–
Trichostrongylus colubriformis	Guinea pig	+	+	+	+	+
Aspiculuris tetraptera	Mouse	+	?	?	–	–
Strongyloides ratti	Rat	+	+	+	?	+

Taken from Wakelin (1978)

intake of a large number of infective juveniles. This may be a response to the antigenic stimulation provided by the exsheathing fluid produced by the juveniles as they complete their moult (see Chapter 6).

Inflammation of the intestine and high levels of IgE in the blood and histamine in the mucosa suggest that immediate hypersensitive reactions are responsible for worm expulsion from the gut. The expulsion of *Nippostrongylus* from the rat is accompanied by a rise in the numbers of mucosal mast cells. In some strains of rat, however, worm expulsion occurs before the increase in mast cells, and in lactating rats, mast cell numbers rise without expulsion occurring (Wakelin, 1984a). Although immediate hypersensitivity appears to be involved in *Nippostrongylus* infections, it may not be essential for expulsion. However, these studies have relied upon histological techniques to determine the numbers of mast cells present in the mucosa. This can give no indication of the rate of turnover of the mast cells or of their activity. The levels of mast cell protease in the blood correlate well with worm expulsion, even though relatively low mast cell numbers can be demonstrated histologically (Wakelin, 1984a). There is more direct evidence for the involvement of hypersensitive/inflammatory responses in the expulsion of *Trichinella spiralis* from the mouse and rat and of *Trichostrongylus colubriformis* from the guinea pig. In these species there is a close correlation between worm expulsion and mast cell numbers and inflammatory changes in the gut wall.

A variety of reactions may be involved in worm expulsion from the gut (Figure 5.6). Mast cell and basophil degranulation releases amines which increase the permeability of the intestinal wall and might attack the nematode directly. The increased permeability allows the passage of IgG and IgE, which normally cannot traverse the intestinal mucosa, in addition to IgA and IgM. Increased vascular permeability facilitates the accumulation of mast cells and other effector cells (notably eosinophils), which leads to the development of a local inflammatory response. A major feature of this inflammation is a large increase in mucus production from the goblet cells. This is accompanied by atrophy of the intestinal villi and increased peristalsis. These changes result in an environment inhospitable to the nematode and expulsion follows. The role of antibodies in this response might be to aid the adhesion of mucus to the surface of the worms, or they may interfere with feeding and reproduction by acting against excretory/secretory antigens.

Figure 5.6: Complexity of the Mechanisms that May Be Involved in the Host Reaction to Intestinal Nematodes and Other Infections. Antigens pass through the intestinal wall mainly via gut-associated lymphatic tissues (GALT) and induce responses in the lymphocytes. This initiates a complex of immune and inflammatory responses which produce substances active against the parasite or changes in the gut environment detrimental to the parasite. Redrawn from Befus and Bienenstock (1984)

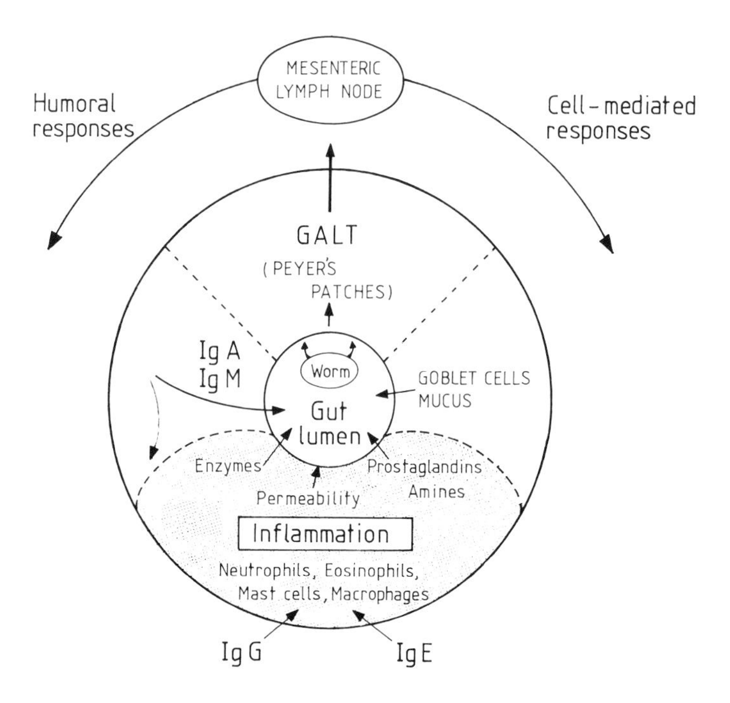

Filarid nematodes infecting the blood and lymphatic systems are more directly accessible to the immune system. Surface antigens are the most important in eliciting an immune response, and antibody production against cuticular antigens has been demonstrated (Wakelin, 1984a). Antibody-mediated cell adherence to the surface of microfilariae can result in their death, and microfilariae in the blood may be suppressed following several months' infection. This response is mediated by IgM in association with neutrophils and macrophages (*Dipetalonema vitae* in hamsters), IgE and eosinophils (*D. vitae* in rats) or IgG and eosinophils (*Wuchereria bancrofti* and *Onchocerca volvulus* in humans) (Wakelin, 1984a).

Immune Evasion

The long-term persistence of many nematode infections implies that they fail to elicit or can survive the immune response of the host. The tough collagenous cuticle may make intestinal nematodes physically and chemically more resistant to immune responses than are other parasites that possess a more naked tegument. Nematodes may also be able to actively evade the immune response. Several evasion mechanisms have been suggested but there is, as yet, no conclusive evidence in support of these.

A general suppression of the immune system is associated with some nematode infections. *Nematospiroides dubius* can establish chronic infections which will persist for several months. Infection with irradiated juveniles elicits a strong protective immunity to subsequent infection. If normal juveniles or adults are present, the response is suppressed, and if the adult worms are transplanted directly into the intestine, subsequent exposure to irradiated juveniles fails to stimulate protective immunity. These observations suggest that the adult worms suppress the immune response of the host (Behnke, Hannah and Pritchard, 1983). This response is non-specific and the immune response to concomitant infections with other helminths is also affected. The mechanisms of this immunosuppression are unknown.

Nematodes may have been able to mask or disguise their antigens by evolving components similar to those of host origin and thus not be recognised by the host's immune system (McLaren, 1984). *Ascaris* collagen, for example, shows cross-reactivity with human collagen but not with mouse or guinea-pig collagen. Alternatively, surface antigens may be masked by absorbing antigens of host origin on to the surface of the nematode. A variety of host-derived proteins have been demonstrated on the surfaces of microfilariae. These include A and B blood-group determinants, immunoglobulins and host serum albumin (McLaren, 1984). However, evidence that the absorption of host molecules provides a functional disguise for the microfilariae is lacking.

Labelling of the surface antigens of *Trichinella spiralis* indicates that surface components are continually shed and replaced (Ogilvie, Philipp, Jungery, Maizels, Worms and Parkhouse, 1980). A rapid turnover of surface components may prevent the adherence of effector cells to the surface of the nematode.

Hosts vary in their susceptibility to parasites. Different strains of mice show different speeds of response to *T. spiralis* (slow responders and rapid responders). These differences are inherited and are thus under genetic control. Crosses between rapid and slow responders resulted in all the F_1 generation being rapid responders, indicating a simple dominant inheritance. A parasite population may be able to evade the host response by surviving in susceptible individuals within the host population (Wakelin, 1984b).

Invertebrate Immunity

Although the defence reactions of invertebrates are not as well understood as those of vertebrates, it is clear that most invertebrates can recognise 'non-self' and mount some sort of reaction. The defence reactions of insects to nematode, and other, infections have been the most studied (Poinar, 1983). Cellular defence reactions result in the blood cells (haemocytes) surrounding and eventually killing the nematode in a process known as encapsulation. This may be simply an accumulation of haemocytes on the surface of the nematode, or it may involve the production of melanin which causes the capsule to brown and harden, forming a solid wall around the parasite. Melanin from the blood can also directly attack the surface of the parasite. Similar reactions have been described in other invertebrates but they are poorly understood.

Plant Parasitic Nematodes

Plant hosts also possess defences against nematodes and other pathogens. Although analogies have been drawn between the plant's defence system and the vertebrate immune system, the response is not as specific as an immune response. It is perhaps more appropriate to think in terms of plants being susceptible or resistant to nematode infections. The host ranges of plant parasitic nematodes vary. Some will infect a variety of hosts whereas others have very restricted ranges. In those species with wide ranges, the population may be divided into races restricted to particular host species. The nematode is adapted to respond to the physiological cues and to circumvent the defences of a restricted range of host species.

Passive resistance involves physical barriers to infection and the production of chemicals that are toxic to nematodes (Wallace, 1973). The resistant host may fail to provide stimuli essential for the development of the nematode. Some plant nematodes hatch in response to host root exudates. This stimulus can be very specific, and hatching will not be elicited by an inappropriate host. Plants may produce chemicals that are toxic to nematodes and other pathogens (allelochemics), such as terthienyls produced by marigolds (*Tagetes* spp.) and nimbidin and thionemone produced by margosa (*Azadirachata indica*) (Veech, 1981).

The plant may respond more actively to the infection in a way that kills the nematode or limits its spread. Structural changes such as lignification, and biochemical changes resulting in necrosis, may physically isolate the nematode. The pattern of nutrient flow may also be altered, depriving the nematode of food. These responses have been collectively called a hypersensitive reaction (Veech, 1981).

The series of biochemical changes resulting in a hypersensitive reaction are unknown. Plant nematodes secrete a variety of enzymes into the host cell including cellulases, proteases and carbohydrases (Giebel, 1982). These enzymes may induce a hypersensitive reaction by releasing host chemicals stored in an inactive form or by inhibiting host regulatory systems. A number of host chemicals have been implicated in this response, including indoleacetic acid (IAA) and other plant hormones, amino acids and phenolic compounds. Chemicals that have a direct nematocidal effect (phytoalexins) have been found (Veech, 1981). Coumesterol is produced by lima beans resistant to *Pratylenchus scribneri*, and glyceollin by soybean resistant to *Meloidogyne incognita.* These phytoalexins can be found throughout the plant but accumulate at the site of infection in concentrations sufficient to cause the death of the nematode.

Giebel and co-workers have proposed what has been called a holistic mechanism for the resistance of potato cultivars to *Globodera rostochiensis* (Giebel, 1982). They suggest that the nematode secretes a β-glycosidase which releases free phenol stored in the plant as a phenolic glucoside, and this directly or indirectly results in necrosis. In susceptible plants the phenol inhibits indoleacetic acid hydrolase, resulting in increased IAA activity and giant cell formation. In resistant plants IAA hydrolase is not inhibited and the phenol is metabolised to lignin, which is involved in necrosis and the hypersensitive reaction (Figure 5.7). This hypothesis is, however, unproven. Many of these events happen some time after resistance becomes apparent, and there is some doubt as to whether the nematode does actually release a β-glycosidase (Veech, 1981).

Figure 5.7: The Complex of Factors which Have Been Suggested to Be Involved in the Holistic Response of Potato Cultivars to *G. rostochiensis* (see text)

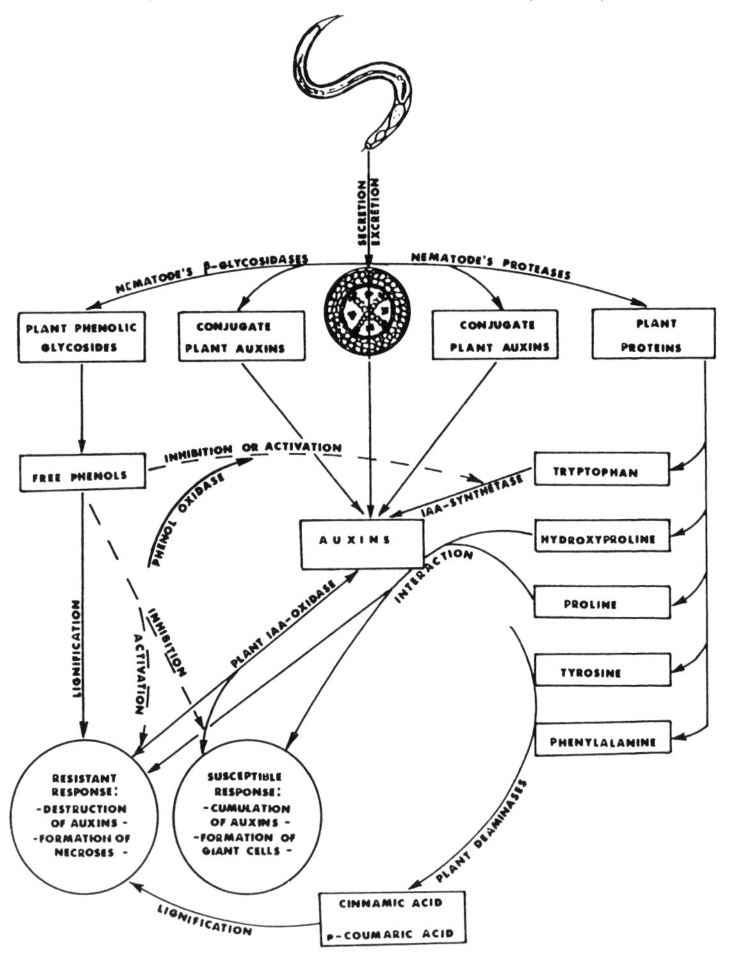

Reproduced from Giebel (1982) with permission

Control

Nematodes cause many economically important diseases of humans, domestic stock and plant crops. Much effort has therefore been expended in an attempt to control these diseases. A variety of control measures have been used including manipulating the environment to the detriment of the parasite, vector control, biological control, and vaccination. So far, the most successful control measures have proved to be the use of drugs and the breeding of nematode-resistant hosts.

The development of drugs against nematodes (confusingly called 'anthelmintics' when applied to animal parasites and 'nematicides' when applied to plant parasites) has been by random or semi-random screening. A wide spectrum of chemicals is applied to a laboratory infection in the hope that one will be found that kills the parasite without killing the host. A more rational approach is to examine differences in the biochemistry of the parasite and the host. This will identify target sites for chemotherapy which will be harmful to the parasite but not to the host. The rational approach has yet to produce a successful drug, but, as Barrett (1981) has pointed out, if as much effort had been put into this approach as has been expended on random screening, perhaps it would have had more success.

A wide range of anthelmintics have been developed, mainly for use against trichostrongyle nematodes of sheep and cattle (Prichard, 1978). The most widely used of these are various benzimadazole-derivatives and levamisole. The mode of action of a number of anthelmintics has been listed by Barrett (1981). Most act by affecting the neuromuscular system or by disrupting the nematode's energy metabolism.

Nematicides active against plant nematodes include halogenated aliphatic hydrocarbons (e.g. methyl bromide) and methyl isothiocyanate mixtures used as soil fumigants. These kill a wide range of soil organisms and are thus of limited usefulness. Organophosphates and oximecarbamates (e.g. aldicarb and oxamyl) impair neuromuscular activity, affecting feeding, invasion and other behavioural activities (Southey, 1978; Wright, 1981). Little is known about the biochemistry of plant parasitic nematodes, and a rational approach to the chemical control of plant parasitic nematodes has not been explored. Promising target sites include the nervous system, energy metabolism and the synthesis of cuticle collagen and egg-shell chitin (Wright, 1981).

The development of drug resistance is a major problem with chemical control. Benzimadazole and levamisole resistance has appeared in trichostrongyle infections of sheep in Australia (Le Jambre, 1978) and more recently in Europe. Drug resistance promises to be a major problem and makes the identification of new target sites for drug action a matter of urgency.

The use of nematode-resistant plant varieties is often the only economic control available against plant nematodes on a field scale. Resistant varieties are developed by breeding and by crossing with wild resistant strains (Southey, 1978). Although the susceptibility of vertebrate hosts to infection varies and susceptibility/resistance is inherited, the breeding of resistant hosts has not been used for the control of animal parasitic nematodes but is a promising area for research.

Host/Parasite Population Dynamics

The application of the theory of population dynamics to host/parasite systems in recent years has added much to our understanding of the nature of parasitism. Parasitism can be considered as an interaction between two populations, the host population and the parasite population, rather than as a relationship between an individual host and its parasites. Parasitism acts in a similar way to a predator/prey relationship, inflicting mortality on heavily infected hosts and thus acting as a density-dependent regulator of the host population (Anderson, 1978).

Various features of a host/parasite relationship can be quantified. Difficulties with qualitative definitions of parasitism led Crofton (1971b) to produce a quantitative definition in what has become a key paper in parasite population dynamics. The distribution of parasites within a host population can be described mathematically by a probability distribution (Figure 5.8). If the parasites were distributed randomly among the hosts, the data would fit the Poisson distribution. In fact the parasites are heavily overdispersed ($s^2 > \bar{x}$); that is, the distribution of parasites among hosts is much greater than would be expected if they were distributed at random. Various mathematical descriptions are available for overdispersed distributions, the negative binomial distribution being most commonly applied to host/parasite systems. The majority of host/parasite systems that have been examined conform to a negative binomial distribution, including several species of nematodes (Keymer, 1982).

The clumping of parasites within hosts results in an overdispersed distribution; a few hosts are heavily infected whereas the majority of hosts contain few parasites. This increases the

Figure 5.8: The Distribution of the Nematode *Hammerschmidtiella diesingi* among its Host, the Cockroach *Periplaneta americana*. The closed histogram shows the observed distribution and the open histogram that predicted by the negative binomial. Original

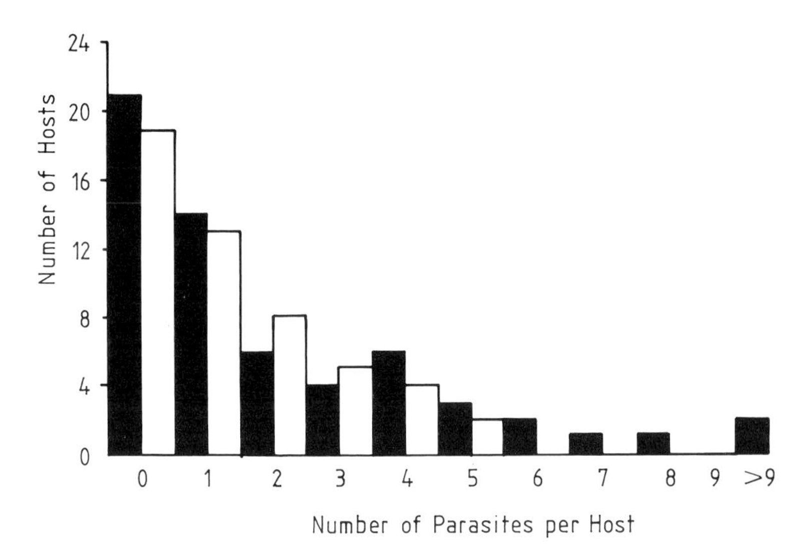

efficiency of density-dependent regulatory processes and hence the overall stability of the host/parasite system. With a direct lethal effect, for example, the death of heavily infected hosts removes a significant proportion of the parasite population. This provides a negative feedback mechanism, regulating both host and parasite populations (Figure 5.9). Other density-dependent restraints are known to regulate nematode populations (Keymer, 1982). Fecundity is inversely related to worm burden (Figure 5.10), and the effect of the immune system is dependent on the size of the worm burden.

A number of workers, notably R.M. Anderson and R.M. May, have formulated mathematical models of host/parasite systems (e.g. Anderson and May, 1978; May and Anderson, 1978). These models have a number of aims including the investigation of factors responsible for stability/instability in host/parasite systems and the effect of control measures. Anderson (1982) has described a model applied to nematodes with a direct life cycle, such as *A. lumbricoides* and hookworms (Figure 5.11). The model consists of a series of coupled differential equations describing the birth (or immigration) and the death (or emigration) rates of the parasite

Figure 5.9: Diagram Showing how a Lethal Effect Imposed by Parasites upon Hosts Can Act as a Negative Feedback Mechanism which Regulates both Populations. Original

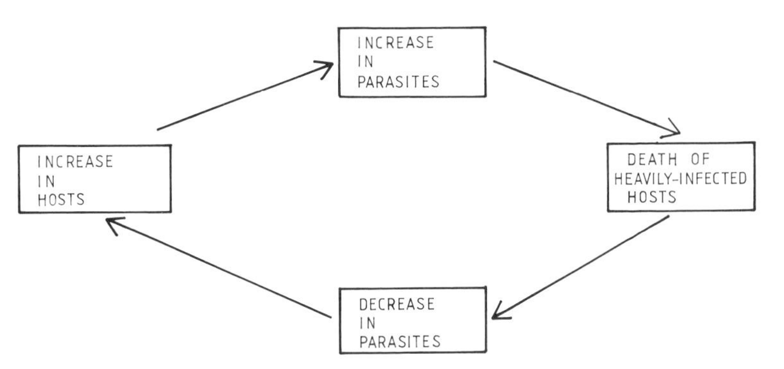

Figure 5.10: The Effect of Population Density on the Fecundity of the Dog Hookworm, *Ancylostoma caninum*. The graph shows the relationship between egg production per female worm (eggs per gram of faeces) and the number of worms within the host. Redrawn from Krupp (1961)

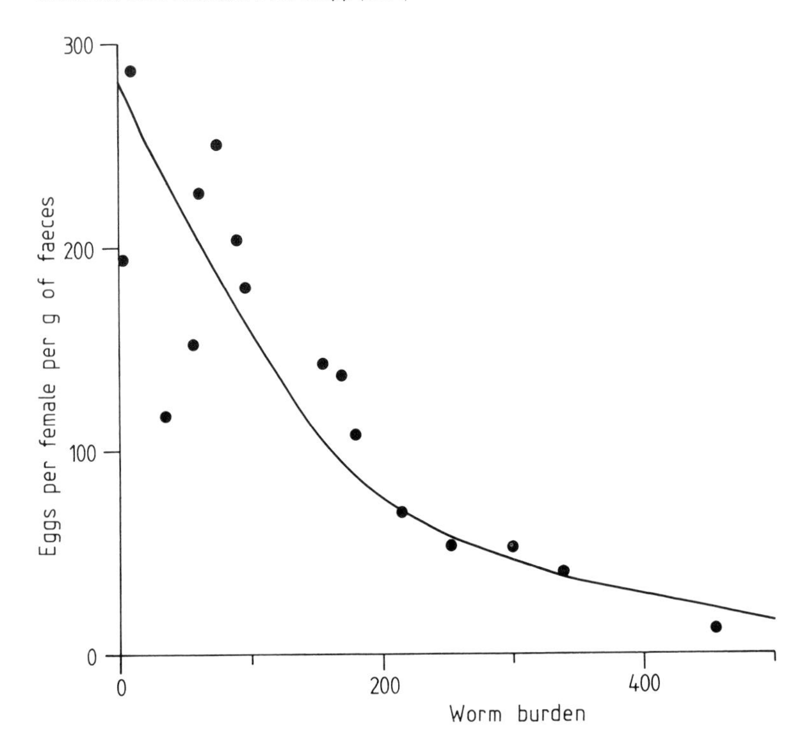

Figure 5.11: Diagram Showing the Organisation of the Model Proposed by Anderson (1982) for a Nematode Parasite with a Direct Life Cycle. It shows the principal populations and rate processes involved. Redrawn from Anderson (1982)

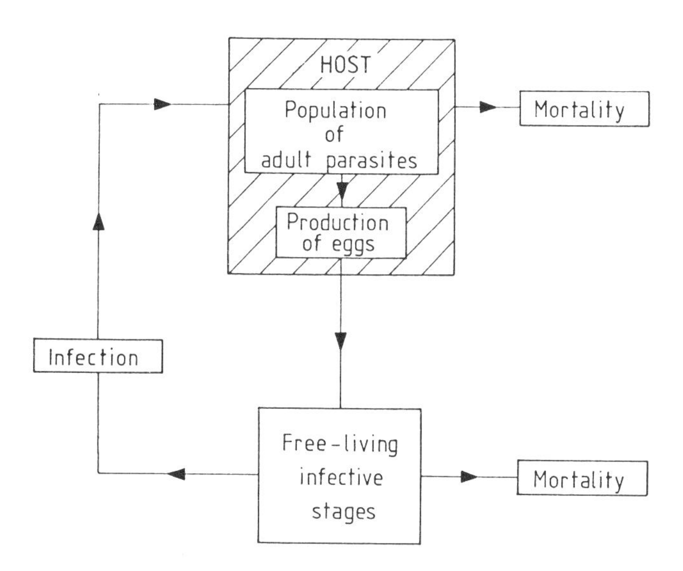

subpopulations (adults, eggs and juveniles). In their simplest form the dynamics are described by the following equation:

$$\frac{dN_t}{dt} = \lambda - (\mu N_t) \tag{5.1}$$

Where dN_t/dt is the rate of change in parasite numbers (N_t) at time t, λ is the birth or immigration rate, and μ is the death or emigration rate. The model is complicated by the dynamics of the host population, time delays in the system and, in the case of amphimictic nematodes, the probability of finding a mate.

The high fecundity of nematode parasites, density-dependent constraints and the overdispersed distribution all tend to increase the stability of the relationship. It is therefore difficult to perturb the system in such a way that disease eradication will ensue. Anderson (1982) suggests that the most appropriate control measures for hookworms and ascariasis are to concentrate chemotherapy on heavily infected individuals (taking advantage of overdispersion) and to improve sanitation to reduce the rate of transmission.

6 LIFE CYCLE

The Basic Pattern

The life cycle of nematodes consists of six stages or instars: the egg (or embryo), four juvenile stages (J1, J2, J3, J4) and the adult (Figure 6.1). All nematodes possess this basic pattern. The immature stages have been called 'juveniles' by plant nematologists and 'larvae' by animal helminthologists. The use of the term 'juvenile' has become more widespread recently. The objection to the use of 'larvae' is that, by analogy with insects, a larva is an immature stage which undergoes a marked metamorphosis into the adult. The immature stages of nematodes are superficially similar to the adult. At an ultrastructural level, however, marked changes may be seen between one stage and the next; for example in the structure of the cuticle (Bird, 1971). In spite of this, I concur with the use of the term 'juvenile' in the interest of maintaining a common terminology.

In discussing the variety of nematode life cycles it may be useful to adopt some of the terminology for insects. In both nematodes and insects each instar in the life cycle is separated by a moult (with the exception of the egg — J1). Three phases are recognised in the moulting cycle of insects: apolysis (the separation of the cuticle from the underlying hypodermis), cuticle formation and, finally, ecdysis — the shedding of the remains of the old cuticle (Chapman, 1972). A similar sequence of events occurs in nematodes.

During the early 1970s there was some controversy among entomologists over whether the end of one instar and the beginning of the next was marked by apolysis or ecdysis. According to Hinton (1973), an instar went from apolysis to apolysis, and a stage that was covered by an apolysed, but not yet ecdysed, cuticle was said to be pharate (hidden).

A similar terminology has recently been applied to nematodes (Rogers and Petronijevic, 1982), although they included cuticle formation in the process of apolysis. In insects the cuticle is opaque and thus, although further development may have occurred, the adult may have the appearance of a pupa because of the retention

Figure 6.1: The Basic Life-cycle Pattern of Nematodes. The embryo develops to a 1st-stage juvenile within the egg, hatches and undergoes four moults before developing into the sexually mature adult. Original

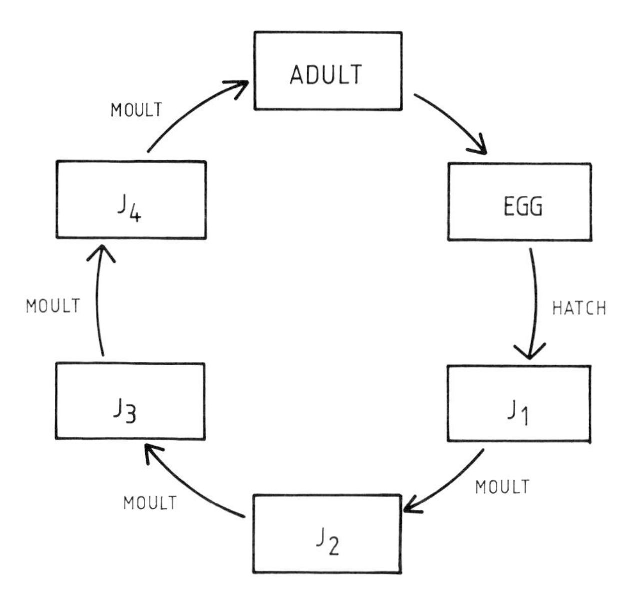

of the pupal cuticle (pharate adult). The nematode cuticle is transparent and, as far as is known, there are no connections between the old and new cuticles. It is, therefore, clear that apolysis marks the beginning of the next instar (Figure 6.2). This terminology is most useful in the interpretation of life cycles where ecdysis or hatching is delayed (Figure 6.2). Recognising apolysis and ecdysis as distinct events aids the interpretation of the control processes involved (Rogers and Petronijevic, 1982).

The Growth Curve and the Function of Moulting

The life cycle of nematodes is punctuated by moulting events, growth and differentiation occurring during the intermoult period. Disagreement about the relationship between moulting and the growth curve of nematodes may be due to differences in the way in which the measurements were made. Byerley, Cassada and Russell (1976) measured size distributions at various times after hatching in synchronous cultures of *Caenorhabditis elegans*. There were no

Figure 6.2: The Relationship between Apolysis (A), ecdysis (E), cuticle formation (shaded area) and hatching (H) in nematode life cycles. (A) The basic pattern. (B) Trichostrongyle nematodes. The 3rd-stage juvenile retains the 2nd-stage cuticle as a protective sheath by a delay in the second ecdysis. Ecdysis occurs when the appropriate stimulus is received upon entry into the host (exsheathment). (C) *Ascaris lumbricoides*. The nematode develops to a 2nd-stage juvenile within the egg which hatches and ecdyses on receipt of a stimulus from the host. Hatching and the first ecdysis are delayed. Original

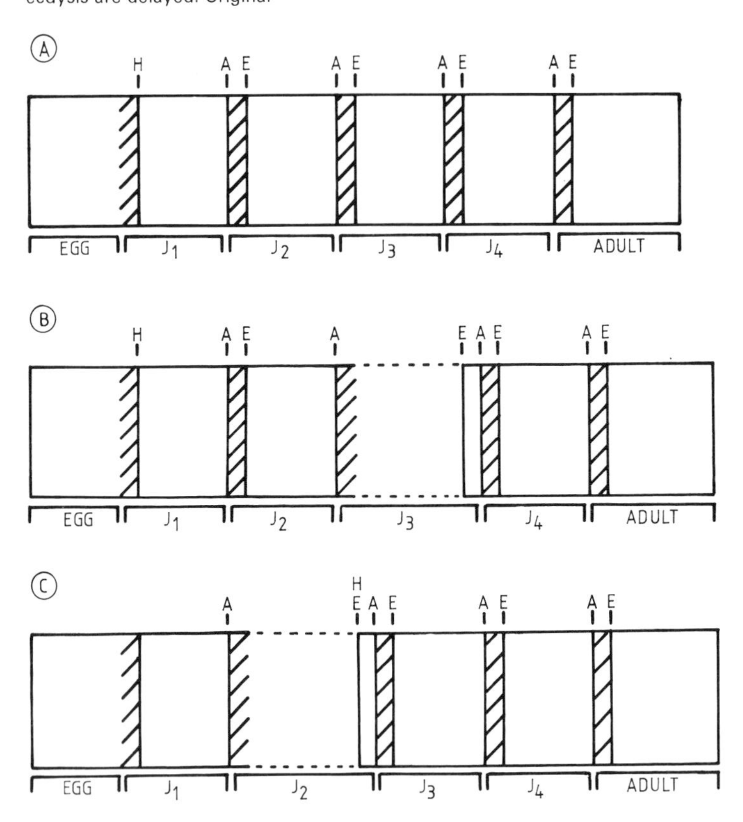

discontinuities in the growth curve, suggesting that moulting had little effect on growth. It is difficult to maintain true synchrony in a nematode culture, and at any given moment the culture contains a distribution of ages and sizes which may iron out any discontinuities in the age/size curve.

Wilson (1976) has attempted to deal with this problem by using asynchronous cultures. This allows simultaneous sampling of all stages in the life cycle. Plotting the number of worms in each size class against length produces a population growth profile with discontinuities that correlate with moulting (Figure 6.3). The overall growth curve of the species consists of the sum of the growth curves of the individual instars. Nevertheless it is clear that significant growth occurs in the intermoult period.

Following the changes in dimension during the development of individual nematodes would provide the best method of determining the growth curve. This is technically difficult and has rarely been attempted. Bird (1983) found that *Rotylenchus reniformis* decreased in volume by 17 per cent during development from 2nd-stage juveniles to immature females and by 19 per cent during development into males. The infective juveniles of *Haemonchus contortus* decrease in volume by 16-30 per cent during exsheathment (Davey and Rogers, 1982). There is apparently no correlation between moulting and growth. This has led Sommerville (1982) to suggest that moulting simply provides an

Figure 6.3: Discontinuities in the Growth Curve of *Caenorhabditis elegans* Revealed in Asynchronous Cultures

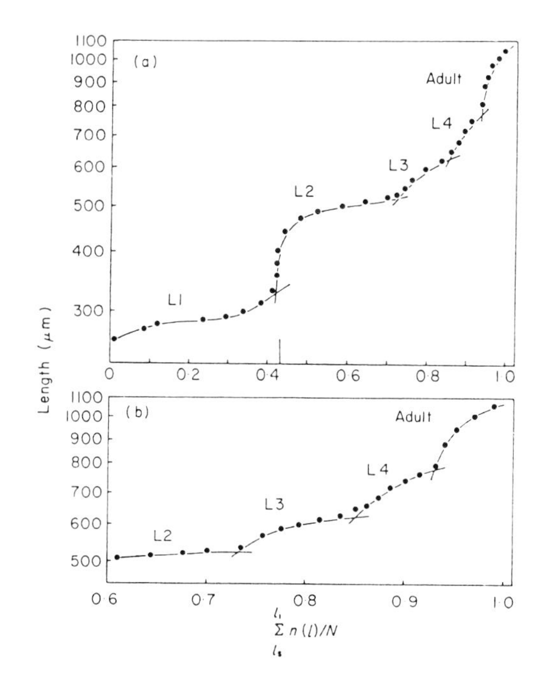

Reproduced from Wilson (1976) with permission

opportunity to replace and modify the cuticle, a method of excreting excess nitrogen, or even that the moulting process is some sort of evolutionary relict.

Nematodes, in particular parasitic species, pass through different environments during completion of their life cycle. Division of the life cycle into instars, separated by moults, allows specialisation. Each stage in the life cycle is adapted to fulfil a particular function (Table 6.1). In trichostrongyle nematodes, for example, the adult and 4th-stage juvenile are parasitic and adapted to the environment within the host. The other instars are free-living. A moult appears to be necessary to accommodate the change in environment. The 3rd-stage juvenile is the infective stage and the most resistant of the free-living stages.

Rogers and Petronijevic (1982) have taken this analysis further. They suggest that each instar possesses its own gene set, producing the structural and other adaptations of the instar concerned. During the life cycle there is a successive activation and inactivation of the gene sets associated with each instar. In the case of the

Table 6.1: The Functions of Instars in the Life Cycle of Trichostrongyle Nematodes

Instar	Environment	Developmental trigger	Survival abilities
Egg	Passes out of host gut and develops to J1 in the faeces	Continuous, laid by female worms	Resistant to host gut enzymes, chemicals, physical damage, desiccation and sub-zero temperatures
J1	Rapid development in the faeces	Hatch in the presence of water	Susceptible to desiccation and other hazards
J2	Rapid development in the faeces	None	Susceptible to desiccation and other hazards
J3	Migrates on to the herbage — infective stage ingested by host	Retains J2 cuticle as a sheath, exsheathes upon ingestion by host in response to CO_2	Resistant to desiccation, sub-zero temperatures and chemicals
J4	Develops within host gut	May be arrested depending upon conditions experienced by J3	Resistant to host reactions, including immune response and physicochemical conditions
Adult	Within host gut	Arrested J4 develops spontaneously or in response to hormonal changes	Resistant to host immune response and physicochemical conditions

infective 3rd-stage juvenile, activation of the gene set of the next instar (J4) is delayed until the appropriate stimulus is received from the host.

Mechanics and Control of Moulting

Cuticle Formation

Moulting involves apolysis, cuticle formation and ecdysis. In addition to the general body cuticle, the cuticular lining of the oesophagus, rectum, sense organs and excretory and reproductive systems are also replaced. The initial event in moulting is the separation of the cuticle from the underlying hypodermis (apolysis). The new cuticle then begins to form beneath it. The hypodermis undergoes ultrastructural changes during cuticle formation which are indicative of protein synthesis (Sommerville, 1982). More direct evidence for hypodermal involvement in cuticle synthesis comes from laser ablation studies in *C. elegans* (Singh and Sulston, 1978). The lateral alae of the adult cuticle are secreted by the lateral hypodermal cells ('seam cells'). If these cells are ablated before formation of the adult cuticle, there is a corresponding gap in the alae. In some species the hypodermis is poorly developed and the cuticle may be secreted by the muscle cells (Sommerville, 1982).

Some authors consider that the cuticle is formed intracellularly by the hypodermis and is part of the hypodermal cells (Inglis, 1983b). In most species, however, the outer hypodermal membrane can be observed between the hypodermis and the developing cuticle. The cuticle is thus an extracellular structure, and the epicuticle on the outer surface of the cuticle cannot, therefore, be a hypodermal membrane. The apparent absence of a hypodermal membrane between the cuticle and the hypodermis in some species may be due to difficulties in preservation and to the high metabolic activity of the hypodermis during cuticle formation. The new cuticle may be highly folded, allowing a considerable amount of growth during the intermoult period (Howells and Blainey, 1983).

Exsheathment of Trichostrongyle Nematodes

The infective 3rd-stage juvenile of trichostrongyle nematodes retains the cuticle of the previous stage as a sheath. This is an apolysed but not yet ecdysed cuticle. The ensheathed juvenile is non-feeding and resistant to environmental extremes and may survive on the pasture for several months. After ingestion by the host the moult is completed by the shedding of the sheath. This is known as exsheathment (actually an ecdysis). Exsheathment is induced by a specific stimulus from the host. This consists of dissolved carbon dioxide (Figure 6.4) with optimum conditions of temperature, pH, redox potential and gut enzymes varying from species to species (Rogers, 1960). *Dictyocaulus* spp. may require exposure to acid pepsin (Lackie, 1975).

The formation of a hyaline ring at the anterior end of the sheath is the first indication of exsheathment. The tip of the sheath lifts off and the juvenile wriggles free. This process can be very rapid (less than 10 min), indicating that the substances responsible for exsheathment are stored in an inactive form rather than synthesised *de novo*.

Leucine aminopeptidase, a lipase and a collagenase have been found in the fluid produced during the exsheathment of *H. contortus* infective juveniles (Rogers, 1982). Purified extract of the collagenase is the only enzyme that shows activity against isolated sheaths and is the most likely candidate for the exsheathing enzyme. Leucine aminopeptidase was thought to be responsible for exsheathment but it shows no activity against sheaths. This enzyme may be responsible for the release of the exsheathing enzyme stored in an inactive form, by causing membrane breakdown (Rogers, 1982).

The oesophagus and the excretory cells have both been sug-

Figure 6.4: The Effect of Dissolved Carbon Dioxide Concentration on Exsheathment of *H. contortus* Infective Juveniles. Bicarbonate-carbon dioxide buffer (pH 6.0, 37°C) containing 0.02M sodium dithionite and 0.5M sodium chloride. Redrawn from Rogers (1960)

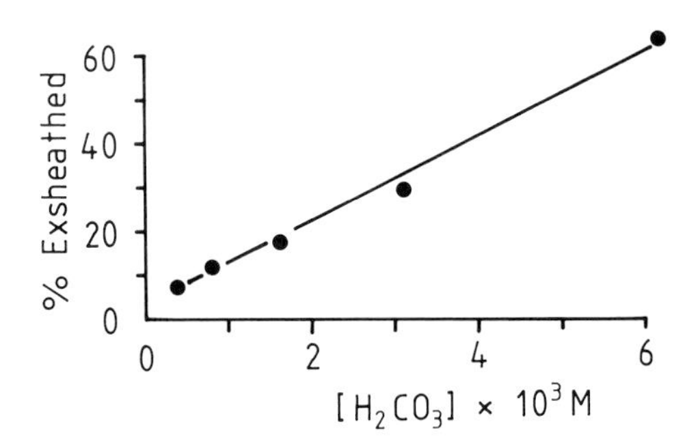

gested as the source of the exsheathing fluid. Both of these structures lose water during exsheathment (Davey and Sommerville, 1982). The excretory cells are tightly packed with electron-dense granules (Plate 2.4), which may represent stored exsheathing enzymes.

The Control of Moulting

Although it is by no means proven, a number of lines of evidence indicate that moulting is under endocrine control (Davey, 1982). Most of this evidence comes from the exsheathment of *H. contortus* infective juveniles and the J4/adult moult of *Phocanema decipiens.* Neurosecretory cells have been identified in *P. decipiens* and these exhibit a cycle of staining during the moult (Davey, 1982). Possible neurosecretory cells have also been observed in *H. contortus* (Rogers and Petronijevic, 1982; Wharton and Sommerville, 1984). Noradrenalin increases during exsheathment and substances possessing juvenile hormone activity have also been identified. Ecdysone-like material has been found in a variety of nematodes and identified as ecdysone, 20-hydroxyecdysone and 20,26-dihydroxyecdysone in *Dirofilaria immitis* (Mendis, Rose, Rees and Goodwin, 1983). Juvenile hormone mimics inhibit hatching and other developmental processes.

Rogers and Petronijevic (1982) have suggested a central role for juvenile hormone in both the control of exsheathment and the development of the 4th-stage juvenile (Figure 6.5). The involvement of a carbonic anhydrase-mediated receptor is indicated by the stimulatory effect of carbon dioxide and the inhibition of exsheathment by carbonic anhydrase inhibitors.

This hypothesis remains unproven. The tissues responsible for hormone production have yet to be identified, and activity that can be related to the observed biological events remains to be demonstrated. The concentration of juvenile hormone that has inhibitory effects *in vitro* is much greater than would be expected if it was acting as a hormone, and may simply provide a non-specific stress (Willett, 1980).

Dauer Larvae

Bacterial-feeding rhabditid nematodes reproduce continuously when conditions are favourable. When conditions become

Figure 6.5: The Control Mechanism for Exsheathment in Trichostrongyle Infective Juveniles, Suggested by Rogers and Petronijevic (1982). The stimulatory effect of carbon dioxide on a postulated carbonic anhydrase-mediated receptor allows the juvenile to escape the inhibitory activity of juvenile hormone. This results in the removal of gene inhibition, allowing the development of the parasitic 4th-stage juvenile, and stimulates the release of exsheathing fluid via a neurosecretory mechanism. Original

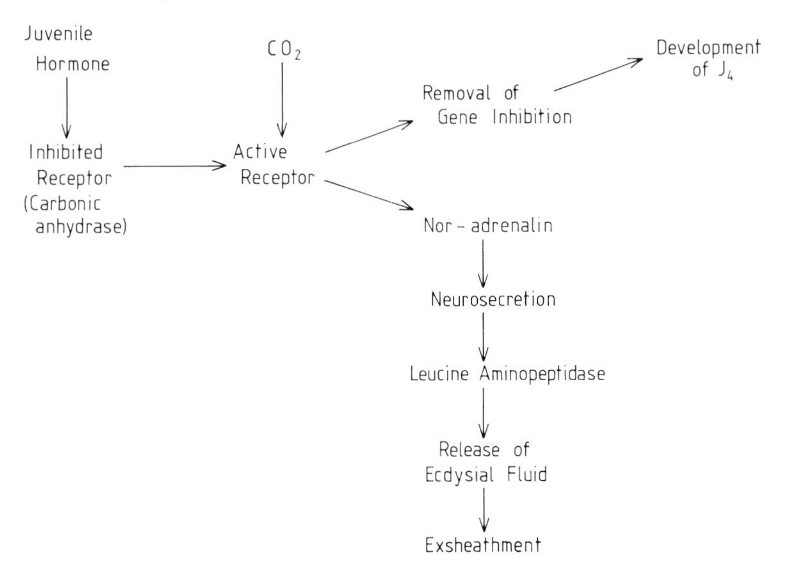

unfavourable, because of desiccation or exhaustion of the food supply, they form a special, resistant, 3rd-stage juvenile called a dauer larva. Dauer larvae are also formed by some insect parasitic nematodes.

The dauer larva of *C. elegans* appears to form in response to gradual starvation, crowding or the accumulation of an inhibitory substance (Cassada and Russell, 1975). It is much thinner than a normal 3rd-stage juvenile, possesses a modified cuticle and does not feed, showing no oesophageal pumping. The epicuticle is much thicker in the dauer larva and the basal zone has a striated layer which is absent in all other stages. These differences in cuticular structure are related to a decrease in the permeability of the cuticle. Dauer larvae are more resistant to desiccation and to chemicals that are lethal to other stages. They also exhibit different behavioural patterns from normal 3rd-stage juveniles.

The dauer larvae are normally inactive but exhibit a burst of activity if stimulated. They will mount projections on the substrate, stand on their tails and wave their heads around. Some dauer larvae attach to the surface of insects, using them as transport hosts to carry them to fresh substrates (phoresis). The behaviour of the dauer larva enables it to locate such a host.

Dauer larva formation in *C. elegans* has aroused the interest of developmental biologists. Normal 3rd-stage/dauer larva formation is a developmental choice which is amenable to experimental investigation, particularly given the ease of mutant isolation in this species. Mutants with defects in dauer larva formation can be isolated by exploiting the fact that dauer larvae are resistant to detergents such as sodium dodecyl sulphate. Two classes of mutant have been isolated. Constitutive mutants produce dauer larvae even when food is available, and defective mutants fail to produce dauer larvae upon starvation (Riddle, 1980). Dauer-defective mutants show structural defects in their amphidial sense organs. This suggests that there is a sensory trigger for dauer larva formation.

Hatching

In most species, hatching occurs as soon as development of the 1st-stage juvenile is complete, providing that water is present. The infective juvenile of some parasitic species, however, develops within the egg and only hatches when it receives an appropriate stimulus from the host. *Ascaris* develops to a 2nd-stage juvenile within the egg which hatches after ingestion by the host (Figure 6.2). The hatching stimulus is provided by carbon dioxide, in a similar fashion to the stimulus producing exsheathment in trichostrongyles (Rogers, 1960).

The plant parasitic nematode, *Globodera rostochiensis*, develops to a 2nd-stage juvenile within the egg. The eggs are protected by a cyst formed by the body of the female and may survive in the soil for several years. Hatching is initiated by some factor present in root diffusates from the potato host. Infective juveniles are thus released when a suitable host is available for infection. The identity of the substance in potato root diffusate which stimulates hatching in *G. rostochiensis* is unknown but a factor causing hatching in *Heterodera glycines* has been isolated from kidney bean roots and designated as glycinoeclepin A (Masamune, Anetai, Takasugi and Katsui, 1982). Studies on the hatching mechanisms of nematodes have concentrated on *A. lumbricoides* and *G. rostochiensis*. The mechanism is thought to be similar in all species, although the hatching stimulus varies (Perry and Clarke, 1981).

Perry and Clarke (1981) have suggested that an increase in the permeability of the egg-shell plays a central role in the hatching process of nematodes. Increased permeability is indicated by the leakage of trehalose from the perivitelline fluid into the surrounding medium. This is accompanied by an increase in the water content of the infective juvenile as the osmotic concentration within the egg decreases, and results in the activation of the infective juvenile. The hatching stimulus may alter egg-shell permeability by acting directly on the lipid layer or indirectly by stimulating the juvenile to release hatching enzymes.

Protease, lipase, chitinase, α-glucosidase and β-glucosidase activity have been demonstrated in the hatching fluid of *A. lumbricoides* (Barrett, 1976). Hatching fluid does not affect the permeability of the lipid layer. The permeability change must occur before the hatching enzymes can attack the other layers of the egg-shell. Barrett (1976) could not find any chemical or conformational changes in the lipid layer of *A. lumbricoides* which could be associated with the change in egg-shell permeability during hatching. He suggested that changes were too localised to be detected or that increased permeability resulted from mechanical damage inflicted by the activated juvenile. Clarke and Perry (1983) have demonstrated that Na^+ binds to the egg-shell of *Ascaris suum* in the presence of the hatching stimulus, and suggest that the ascarosides of the lipid layer form tori in the presence of $NaHCO_3$ which provide channels allowing trehalose leakage.

Atkinson and Ballantyne (1979) have suggested a messenger role for calcium in the hatching of *G. rostochiensis* eggs but this is disputed (Perry and Clarke, 1981). Atkinson and Taylor (1980) have postulated a sialoglycoprotein calcium binding site on a membrane associated with the lipid layer. The presence of lipoprotein membranes in the lipid layer has now been demonstrated (Perry, Wharton and Clarke, 1982), and Clarke and Perry (in press) have provided evidence of calcium in these membranes and of its displacement by hatching agents. Whether calcium has a messenger role or a structural function remains to be established, but the result of treatment with hatching agents is an alteration in the permeability of the egg-shell, allowing the release of trehalose and the hydration and activation of the juvenile.

The infective juvenile of *G. rostochiensis* escapes from the egg by cutting a slit in the shell with a series of co-ordinated thrusts of its stylet. In some species the juvenile escapes via a specialised region of the egg-shell which forms an operculum (Wharton, 1980). The eggs of trichurid and capillarid nematodes are barrel-shaped with an opercular plug at either pole. The arrangement of chitin and protein in the chitinous layer of the plug differs from that of the rest of the shell. This may make the opercular plug more susceptible to enzymatic attack during hatching (Wharton and Jenkins, 1978).

Diapause Phenomena

Diapause is a delay in development which is not a direct result of adverse environmental conditions and is not immediately ended by a return to favourable conditions (Shelford, 1929). Evans and Perry (1976) have applied this term to interruptions in the development or life cycle of nematodes.

Several phenomena in the life cycle of nematodes could be considered as a diapause. A nematode in diapause does not respond to cues that would otherwise induce further development. This may be a normal part of the life cycle (obligate diapause) or may be initiated by environmental conditions (facultative diapause). The state of diapause may end spontaneously or in response to environmental cues. The nematode will then resume development or respond to developmental triggers (Evans and Perry, 1976). In insects diapause is under hormonal control and we might expect this also to be the case in nematodes, but there is no evidence for this. The relationship of diapause to other forms of dormancy will be considered in Chapter 7.

Arrested Development

In Britain, infective juveniles of trichostrongyle nematodes ingested in the autumn moult to the 4th-stage juvenile and become arrested. They overwinter as the 4th stage within the gut of the host and do not resume development until the following spring. The eggs produced after the maturation of arrested juveniles are responsible for the 'spring rise' in egg production characteristic of trichostrongyle infections.

Arrested development is induced by the conditions experienced by the infective juveniles on the pasture before ingestion by the host. Photoperiod may be involved, but decreasing temperature is the most important stimulus (Schad, 1977). Only a proportion of juveniles become arrested, and a bimodal size distribution in a nematode infection may be an indication that arrested development has occurred. Infective juveniles of *H. contortus* are poor at surviving low temperatures on the pasture and 100 per cent of the overwintering population consists of arrested juveniles.

Apart from providing a means of overwintering, arrested development synchronises the host and parasite life cycles. The spring rise in egg production and the subsequent increase in infective juveniles on the pasture occur just when lambs or calves are starting to feed, increasing the chances of infection. In Australia, juveniles arrest in the spring and survive the hot, dry summer as arrested 4th-stage juveniles within the host (Evans and Perry, 1976). In contrast to British populations, Australian populations of *Ostertagia ostertagi* do not become inhibited after 8 weeks' exposure to 4°C (Smeal and Donald, 1982). The stimulus inducing arrested development is, therefore, different in Australian and British populations. The effect of rising temperatures and increasing photoperiod, simulating Australian spring conditions, has not yet been investigated.

The biochemical and physiological mechanisms underlying the onset and termination of arrested development have yet to be determined. The observation that trichostrongyle infective juveniles are cold tolerant and may possess mechanisms for cold acclimatisation (Wharton, Young and Barrett, 1984) is perhaps of interest. Arrested juveniles may resume development spontaneously after a fixed period of time or they respond to changes associated with the onset of breeding in the host (Michel, 1974).

Egg Diapause

The eggs of *Nematodirus battus* are thought to exhibit a clear diapause (Evans and Perry, 1976). The 3rd-stage infective juvenile develops within the egg. In Britain, eggs that have developed on pasture fail to hatch if kept at 20°C, but chilling at 5°C for 6-8 weeks followed by transfer to 20°C elicits a mass hatch. Conditions in the autumn appear to induce a diapause which is terminated by chilling over the winter, followed by a rise in temperature in the spring. The infective juveniles thus overwinter in the egg and hatch *en masse* in the spring ready to infect newly weaned lambs. This response is so predictable that formulae have been derived which forecast the timing and severity of pasture contamination with hatched juveniles (Thomas, 1974).

During chilling, lipid is converted into carbohydrate (mainly glycogen and trehalose). Trehalose is known to act as a cryoprotectant in cold-tolerant insects. Its production may therefore be related to cold tolerance in *N. battus* or to the termination of diapause (Ash and Atkinson, 1983). The eggs of several plant parasitic nematodes are also thought to exhibit a diapause (Evans and Perry, 1976). This is indicated by an increased hatch after chilling.

Life-cycle Theory

Attempts to apply life-cycle theory to the patterns seen in nematodes have largely been confined to free-living nematodes. The data needed to calculate basic life-cycle parameters are not available for parasitic species and would be difficult to obtain. It may be useful, however, to explore what the calculation of reproductive parameters can tell us about the way in which the life cycles of free-living and parasitic nematodes are organised, within the limitations of the data available.

Rate of Natural Increase

A measure of the reproductive potential of a population can be obtained by calculating the average number of (female) offspring produced per female entering the population (R_0 — the net reproductive rate) and the instantaneous growth rate of the population, assuming a stable age distribution and under conditions where space and resources are not limiting (r — the intrinsic rate of increase). R_0 is calculated from life tables (Poole, 1974) as the sum to the maximum age class reached in the population (t) of the number of female offspring produced per female during each time interval (m_x) times the probability of surviving from birth to that age (l_x):

$$R_0 = \Sigma_0^t \, l_x m_x \tag{6.1}$$

r is most accurately calculated as the intrinsic rate of natural increase (r_m) by the solution of the Lotka equation:

$$\Sigma_0^\infty \, l_x m_x \exp(-r_m x) = 1 \tag{6.2}$$

This also requires the measurement of age-specific survival (l_x) and age-specific fecundity (m_x) summarised in life tables. r is sometimes estimated as the capacity for increase (r_c), calculated from R_0 (equation 6.1) and the cohort generation time (T_c):

$$T_c = (1/R_0) \, \Sigma_0^t \times l_x m_x \tag{6.3}$$

$$r_c = (l_n Ro)/T_c \tag{6.4}$$

r_c (equation 6.4) may then be used to give an initial estimate of r_m, the exact value being obtained by solving equation 6.2 by iteration. Less accurate estimates of r_m have been obtained for nematodes using the minimum generation time T_{min}, which is the minimum developmental time from egg to egg production. The use of T_{min} gives serious underestimates of r_m and should be used with caution (Vranken and Heip, 1983).

Table 6.2 lists the measurements of life-cycle parameters of free-living nematodes. These figures are very dependent upon environmental factors, such as temperature, salinity and food availability. The figures quoted are therefore the maximum observed rates. The r_m values observed for bacteriophagous species which are adapted to an abundant but short-term food supply are among the highest ever recorded for animals in this size range (e.g. *C. briggsae, M. iheriteri*). This is associated with short generation times. Relatively low capacities for increase are found in predacious (e.g. *L. vulvapillatum, O. oxyuris*) and herbivorous (e.g. *E. pararmatus*) species, which have relatively long developmental periods.

Reproductive Strategies in Nematodes

Animals may breed once and then die (semelparity) or breed more than once (iteroparity). Most nematodes are iteroparous and produce gametes continuously after maturation until shortly before death. Exceptions are mermithids, which are non-feeding in the free-living adult phase of the life cycle, and cyst nematodes of the genus *Globodera* and *Heterodera*, where the female dies when

Table 6.2: Life-cycle Parameters of Various Species of Free-living Nematodes

Species	R_0	r_c(day−1)	r_m(day−1)	T_c(days)	T_{min}(days)	Reference
Rhabditis marina	400	0.837	0.914	7.2	4.5	Vranken and Heip, 1983
Mesodiplogaster iheriteri	265	1.015	1.447	5.5	2.3	Vranken and Heip, 1983
Labronema vulvapillatum	155	0.077	0.099	65	31	Vranken and Heip, 1983
Caenorhabditis briggsae	153.6	—	1.136	4.43	3.13	Schiemer, 1982b
Plectus palustris	—	—	0.28	—	15.5	Schiemer, 1983
Oncholaimus oxyuris	—	—	0.029*	—	101	Heip, Smol and Absillis, 1978
Eudiplogaster pararmatus	—	—	0.08*	25.3*	—	Romeyn, Bouwman and Admiraal, 1983
Acrobeloides sp.	40.1	—	0.34*	—	11	Anderson and Coleman, 1981

*Estimates

gravid, the body forming a protective cyst.

The capacity for increase of the population is dependent upon the fecundity of females, the generation time and adult and juvenile survival (equations 6.1 and 6.4). The potential for reproduction can thus be increased by increasing fecundity or juvenile or adult survival, and by decreasing the generation time. Nematodes use all of these strategies (Table 6.3). Juvenile survival is maximised by the presence of resistant stages in the life cycle.

The life-cycle strategies employed by parasitic and free-living nematodes are markedly different. Animal parasitic nematodes tend to have high fecundities and relatively long generation times whereas free-living nematodes tend to have low fecundities and short generation times (Table 6.3). The intrinsic capacity for increase (r_c) is directly related to the logarithm of the fecundity but inversely related to the generation time (equations 6.2 and 6.4). Depending on the initial magnitude of the fecundity, decreasing the generation time results in a greater increase in r_c than would increasing the fecundity (Lewontin, 1965; Meats, 1971; Snell, 1978). The reproductive potential of a free-living species may, therefore, be much greater than that of a parasitic species in spite of a much lower fecundity. *Neoaplectana glaseri* is a bacteriophagous parasite of insects which can multiply within its host. Its life cycle thus resembles that of a free-living species. Crofton (1966) has calculated that *N. glaseri* with a total egg production of 15 eggs/female and a generation time of 4 days has a much greater capacity for increase than *H. contortus* with a fecundity of 5000 eggs/female/day and a generation time of 35 days.

The high intrinsic rate of increase of some free-living bacteriophagous nematodes enables them to rapidly colonise a substrate. This is achieved by having a short generation time. A number of mechanisms found in free-living nematodes ensure a rapid rate of reproduction and development under favourable conditions. Parthenogenetic species or self-fertilising hermaphrodites avoid delays in mate location and copulation. Some species are ovoviviparous and the juveniles hatch from the eggs within the uterus before they are released. Where the young are released from the female at the egg stage, the eggs are laid at an advanced stage of development. This speeds up development by telescoping the life cycle. The maximum rate of increase can, of course, only be maintained over a relatively short period of time. The food supply rapidly becomes exhausted and the nematode may switch to the

Table 6.3: Some Nematode Life-cycle Parameters

Species	Habitat	Fecundity (eggs/female/day)	Breeding	Longevity (adult)	Generation time (days)	Resistant stage	Reference
Ancylostoma duodenale	Animal parasitic	10000-25000	Iteroparous	1 year	69	J3	Hoagland and Schad, 1978
Necator americanus	Animal parasitic	5000-10000	Iteroparous	3-5 years	73	J3	Hoagland and Schad, 1978
Ascaris lumbricoides	Animal parasitic	200000	Iteroparous	1-2 years	80	J2 in egg	Levine, 1968
Haemonchus contortus	Animal parasitic	5000+	Iteroparous	25 days	35	J1 in egg, J3, arrested J4	Levine, 1968
Globodera rostochiensis	Plant parasitic	300 (total)	Semelparous	<1 year	35	J2 in egg and cyst	Various sources
Ditylenchus dipsaci	Plant parasitic	207-498 (total)	Continuous in host	45-73 days	21	J4	Yuksel, 1960
Mermis nigrescens	Animal parasitic (insect)	14000 (total)	Semelparous	3 years	1 year	J1 in egg, J2-J4 in host	Poinar, 1979
Neoaplectana glaseri	Animal parasitic (insect)	15 (total)	Iteroparous	?	4	Dauer larva	Poinar, 1979
Mesodiplogaster iheriteri	Free-living bacteria feeder	47 (total)	Iteroparous	11 days	4	Dauer larva	Anderson and Coleman, 1981
Acrobeloides sp.	Free-living bacertia feeder	66 (total)	Iteroparous	18 days	11	Dauer larva	Anderson and Coleman, 1981
Caenorhabditis elegans	Free-living bacteria feeder	280 (total)	Iteroparous	5.5 days	3.5	Dauer larva	Byerley *et al.*, 1976
Plectus palustris	Free-living bacteria feeder	20	Iteroparous	65.5 days	15.5	—	Schiemer, 1983

production of a resting stage (such as a dauer larva) which awaits transfer to a fresh habitat.

Animal parasitic nematodes do not, in general, multiply within their hosts. Transmission to a new host must occur before the life cycle can be completed, and parasitic species usually possess a resting stage which only resumes development upon entry into a suitable host. The difficulties of transmission not only result in a high wastage rate of juveniles but also ensure that development is delayed. Such delays in transmission mean that parasitic nematodes have a much longer generation time than free-living species. The maximum rate of development under laboratory conditions can give a serious underestimate of developmental periods in the field. It is likely, for example, that trichostrongyle nematodes of sheep and cattle complete only one generation per year (Michel, 1974); whereas, in the laboratory, the minimum generation time of *H. contortus* is 25 days (Table 6.3).

It is often stated that the high fecundity of parasites offsets a high wastage rate of free-living stages (low juvenile survival). However, in the nutrient-rich and protected environment within a host, parasites can maximise their reproductive effort (i.e. increase fecundity) without corresponding adverse effects on adult survival (Calow, 1983b). This may have the effect of compensating for low juvenile survival and the long generation time imposed on parasites by the requirements of host-to-host transmission; but the selective pressure for high fecundities is provided by high levels of resource availability.

The capacity for increase can be maximised by increasing juvenile survival as well as by increasing fecundity (equations 6.1 and 6.2). The life cycle of parasitic nematodes either involves an intermediate host or includes resistant or resting stages. The eggs and infective juveniles of many parasitic nematodes are resistant to desiccation, low temperatures, osmotic stress and harmful chemicals. The presence of resistant stages or the use of an intermediate host decreases the wastage rate by increasing juvenile survival.

r and K Selection

If resources are assumed to be unlimited, the intrinsic rate of increase of the population (r) is also defined by the rate of change in the size of the population (N) with time (t):

$$\frac{dN}{dt} = rN \tag{6.5}$$

Resources can only be assumed to be unlimited during the initial phase of colonisation of an unexploited habitat. As resources become depleted, the capacity for increase is decreased and the population size is dependent upon the carrying capacity of the habitat (K), that is, the maximum number of organisms it will support:

$$\frac{dN}{dt} = rN \frac{(1 - N)}{K} \tag{6.6}$$

Species have sometimes been differentiated by whether they maximise fitness by maximising their intrinsic rate of increase (r-selected: maximise r in equation 6.5) or whether they maximise their share of the carrying capacity of the habitat (K-selected: maximise K in equation 6.6). Conditions of r-selection favour reproduction at the expense of adult survival — semelparity, reproduction at a small body size, rapid development, short lifespan and high fecundity. This is usually found in colonising species and species that regularly experience a high mortality due to environmental extremes.

Where predation and competition are the dominant features of the environment, the survival of the parent rather than a high reproductive capacity will be favoured; that is, K-selection occurs. Large experienced parents can compete for limited resources more efficiently than their small, inexperienced offspring. These conditions therefore favour iteroparity, large adult size, delayed breeding, long adult longevity and the production of relatively few, but well protected and provisioned, offspring. Not all species fit strictly into the r-selected or K-selected categories and may show a mixture of r and K influences in the environment (Calow, 1978).

As originally proposed, the theory of r and K selection was framed in terms of population density. r-selected organisms can maintain a higher rate of increase per individual (i.e. fitness) at low population densities; whereas K-selected organisms can maintain a higher rate of increase at high population densities. Sibly and Calow (1985) consider that the classification of life cycles based on density is inadequate because the rate of increase is dependent upon a number of factors, including fecundity, survival and the growth rate of offspring. They conclude that a high reproductive effort is selected for when there is a high juvenile survival (*r*-selected) and that the production of fewer, larger offspring is selected for in environments in which juveniles grow at a slow rate (*K*-selected).

Some free-living bacteriophagous nematodes, such as *C. elegans*, are colonising species, which has clearly favoured the evolution of *r*-selected characters. They have a short developmental period, breed at a small body size, have a short period of repeated breeding (tending towards semelparity), low adult survival and relatively high fecundities. Other bacteriophagous nematodes, such as *Plectus palustris*, inhabit more stable environments and are likely to experience greater competition. This has favoured the evolution of *K*-selected characters with longer developmental periods, delayed reproduction, iteroparity, longer adult survival and lower fecundities (Table 6.3).

The life cycles of parasitic nematodes are more difficult to categorise and show a mixture of *r* and *K* characters (Table 6.3). They have a high fecundity but relatively long adult survival, iteroparity and a long developmental period. A high level of food availability enables them to maintain high levels of fecundity without low adult survival (Calow, 1983b).

Reproductive Energetics

An optimum reproductive strategy would combine a high fecundity with high adult longevity and repeated breeeding (iteroparity) to guard against the loss of a single season's reproductive output (Calow, 1978). Such a strategy is found rarely among living organisms, suggesting that there are costs associated with reproduction. These costs may be expressed as a decrease in adult longevity or in the parent's subsequent reproductive capacity. This is also predicted by the energy budget of a species.

The energy absorbed (A) by an animal is equal to that ingested (C) minus that lost by defecation (F). The absorbed energy may be expended in the production of tissue (P_G for somatic tissue, P_r for reproductive tissue) or as respiratory heat loss (R). The energy budget is in balance:

$$C - F = A = P_G + P_r + R \qquad (6.7)$$

Diverting resources into reproduction (increasing P_r) may occur at the expense of somatic growth and maintenance (decreasing P_G). This may be mitigated by reproductive production being more efficient than somatic production (increased A relative to R) or by reproduction utilising energy stores accumulated by the parent in the pre-reproductive period (Calow, 1983a). Nevertheless, some trade-off between increased reproductive effort and adult survival is expected. The precise form of this trade-off is critical in determining what reproductive strategy has evolved in a particular set of circumstances. A plausible trade-off curve is shown in Figure 6.6. Increasing fecundity results in an accelerating decrease in adult survival. This might occur because the first few gametes are produced from resources in excess of somatic requirements but increasing gamete production occurs progressively from resources essential to adult survival.

What strategy a particular species has evolved depends upon juvenile survival. High juvenile survival favours high fecundity with a correspondingly high adult mortality (favours semelparity). This has been called 'reproductive recklessness' as high reproductive effort occurs at the expense of adult survival (Calow, 1978). Low juvenile survival favours low fecundity and low adult mortality (favours iteroparity). This strategy has been called 'reproductive restraint' and favours the survival of adults which are more likely to survive adverse conditions than are the juveniles.

Some indication of the strategy evolved by a particular species

Figure 6.6: A Plausible Trade-off Curve between Postreproductive Survival (S_a) and Fecundity (n). Survival decreases at an accelerating rate with increasing fecundity. The actual strategy employed may depend upon the survival of juvenile stages. Low juvenile survival favours low fecundity and prolonged adult survival (A), whereas high juvenile survival favours higher fecundity and a correspondingly lower adult survival (B). The straight lines are fitness isoclines with a negative slope which increases as the chances of juvenile survival increase. Redrawn from Calow (1983b)

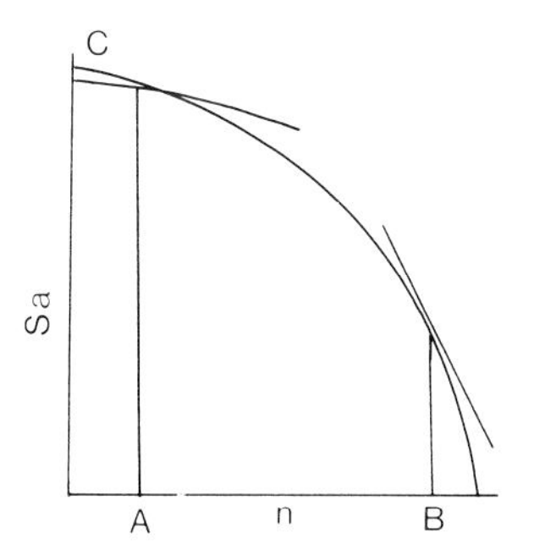

can be obtained by measuring energy budgets and comparing the resources allocated to somatic and reproductive production throughout the life cycle. A measure of the cost of reproduction can be obtained by calculating the cost index, CI (Calow, 1979):

$$\mathrm{CI} = 1 - \frac{A - P_r}{R^*} \qquad (6.8)$$

where R^* is the energy required to support the metabolic demands of the parent's somatic tissues (estimated as the respiratory demand immediately before the start of gamete production). Positive CI values indicate that reproduction occurs at the expense of the maintenance of the somatic tissues, whereas negative CI values indicate that more resources are available for somatic metabolism after than before the start of gamete production.

Relatively complete energy budgets are available for two species of free-living nematodes, *Caenorhabditis briggsae* and *Plectus palustris* (Schiemer, Duncan and Klekowski, 1980; Schiemer, 1982a, 1983). In *C. briggsae* at high levels of food availability (5×10^{10} bacterial cells/ml) respiration increases linearly with age whereas production and absorption increase hyperbolically (Figure 6.7). Although some somatic growth continues after the onset of reproduction, most of the production process is diverted into gamete production. Production efficiency increases markedly during reproduction, this being accommodated by an increase in absorption relative to respiration. The cost index varies from −0.16 to −3.18, again indicating that reproductive production occurs by increasing absorption efficiency rather than at the expense of somatic maintenance.

At lower food availabilities (5×10^9 and 5×10^8 bacterial cells/ml) the period of juvenile development is extended and reproduction occurs at a later age but at a smaller body size. Respiration is maintained at a high level but production is strongly reduced. Although there is some reduction in body size, decreased production is mainly due to a decrease in reproductive output. This is mainly due to a decrease in the numbers of eggs laid but there is also a small but statistically significant reduction in egg size (Schiemer, 1982a). The cost index varies from −0.43 to −1.78, indicating that the reproductive effort is not maintained at the expense of somatic metabolism at reduced food levels. The food

Figure 6.7: Cumulative Energy Budget for *C. briggsae* from Egg Hatching to the End of the Reproductive Period, at Three Different Levels of Food Supply. *A*, Absorption; *R*, respiration; *P*, production; P_G, somatic production; P_r, reproductive production. Data from Schiemer (1982a)

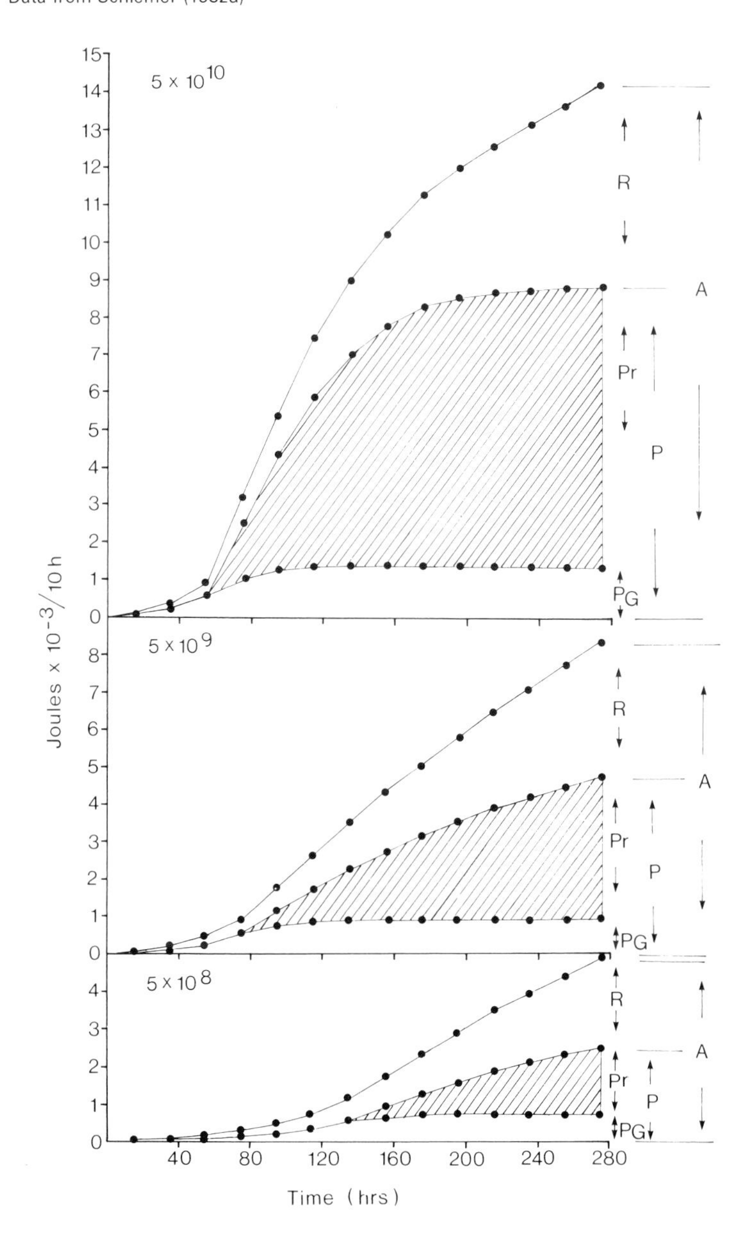

concentration at which juveniles can grow is very close to levels at which they can survive. This indicates a low capacity for the storage of reserve materials and a degree of reproductive recklessness directed towards breeding at an early age (Schiemer, 1983).

Similar conclusions can be drawn from the cumulated energy budget of *P. palustris* (Schiemer *et al.*, 1980). The increased production associated with reproduction is accommodated by an increase in the efficiency of absorption relative to respiration. At high food availability (6-9 $\times$ 10^9 bacterial cells/ml) somatic growth continues after the start of reproduction, in contrast to *C. briggsae.* At reduced food availability (6-9 $\times$ 10^8 bacterial cells/ml) production is reduced and the onset of reproduction is delayed but occurs at a smaller body size. The cost index varies from -0.42 to -1.23, depending upon age and food availability. This again indicates that reproduction does not occur at the expense of somatic maintenance.

The absorption rate of *P. palustris* is approximately one-third that of *C. briggsae,* and its juvenile development time and reproductive phase are much longer (Schiemer, 1983). The food threshold at which growth can occur is much lower in *P. palustris,* but at high food availability *C. briggsae* has a much greater absorption efficiency and capacity for growth (Figure 6.8). The latter is associated with organic pollution and is thus adapted to colonise temporary habitats with high food availability. *Plectus palustris* is adapted to a lower food abundance but perhaps a higher predictability of food supply, and has a greater ability to survive starvation; *C. briggsae,* however, can produce a resistant dauer larva in response to adverse conditions.

Studies on these two species of nematode support the suggestion that reproductive production is more efficient than somatic production (Calow, 1983a). This is achieved by an increase in absorption relative to respiratory losses rather than by gamete production denying resources to somatic maintenance. The energetics and life-history characteristics of *P. palustris* and *C. briggsae* can be correlated with the habitats in which they are found (Schiemer, 1983). Clearly information on a wider range of species from different types of environment is needed to take this analysis further. The ability to manipulate the environmental variables of nematodes in culture makes them attractive subjects for evolutionary and physiological research of this nature.

Figure 6.8: Relationship between the Intrinsic Rate of Increase of the Population (r) and Food Concentration in *P. palustris* (black histograms) and *C. briggsae* (white histograms). The broken vertical line marks the food concentration at which *C. briggsae* becomes superior to *P. palustris*. Redrawn from Schiemer (1983)

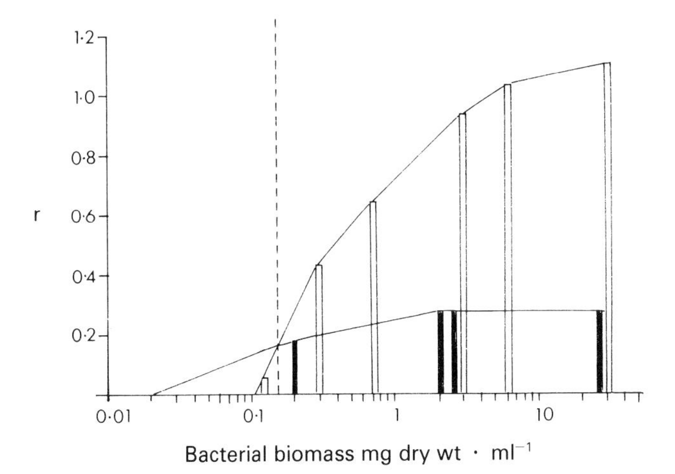

The Parasite Paradox

Parasitic nematodes exhibit high fecundity, low juvenile survival and relatively high adult survival (Table 6.3). If we expect a trade-off between fecundity and adult survival (Figure 6.6), the reproductive patterns observed in parasitic nematodes, and other parasitic helminths, might appear paradoxical (Calow, 1983b). However, as we have seen, the energy budgets of two species of free-living nematodes indicate that the resources for reproductive production in these species come from increased absorption rather than at the expense of somatic maintenance. We might expect similar results from parasitic nematodes. Parasites live in a protected and nutrient-rich environment. The costs of gamete production can, therefore, be met from resources in excess of somatic requirements (Calow, 1983b).

The Size and Number of Eggs

Given a fixed level of resources allocated to gamete production, energy may be channelled into producing more eggs or fewer but larger eggs. Larger eggs are better provisioned than small eggs and may thus have a better chance of reaching maturity (Calow, 1978). There is relatively little variation in the average size of eggs throughout the Nematoda (Bird, 1971) and the size of eggs does not appear to be related to either the fecundity or the size of the adult. Perhaps the mode of oogenesis found in nematodes places limits on the variation in size that is possible.

There is, however, considerable variation in the thickness and complexity of the nematode egg-shell (Wharton, 1980). The most complex egg-shells are found in species where eggs are exposed to environmental hazards, such as desiccation, and particularly where the egg is the stage infective to the host. Nematodes may increase the chances of survival, and thus of infecting a host, by providing a resistant egg-shell rather than by increasing the size of the embryo. Quantitative information on the partitioning of resources into yolk and egg-shell production in relation to fecundity and absorption is needed before conclusions can be drawn about the strategies involved.

Life-cycle Patterns of Parasitic Nematodes

Although retaining the same basic pattern, the life cycles of parasitic nematodes are much more complex than those of their free-living relations. Parasitic nematodes need to get from one host to another, facing the hazards of the environment outside the host as well as the problems of transmission. Different life cycles have evolved that increase the chances of transmission, reduce exposure to the external environment or include stages able to survive such exposure.

Direct Life Cycles

Direct life cycles include only one host, and host/host transmission involves the exposure of free-living stages to the environment outside the host. Different species and stages possess varying degrees of resistance.

The hookworms, *Necator americanus* and *Ancylostoma duodenale*, hatch as 1st-stage juveniles within the soil and develop to an infective 3rd-stage juvenile. Infection occurs by direct penetration through the skin of the host. Although the two species differ, they are both susceptible to environmental hazards such as desiccation (Hoagland and Schad, 1978). This limits their distribution to warm, wet climates such as the tropics.

The 3rd-stage juvenile is also the infective stage of trichostrongyle nematodes, including those infecting sheep and cattle (Table 6.1). Both the 1st-stage juvenile within the egg and the ensheathed infective juvenile are resistant to desiccation, chemicals and low temperatures (Wharton, 1982c; Wharton *et al.*, 1984) and the infective juvenile can survive on pasture for several months before infecting a host.

Ascaris lumbricoides, Globodera rostochiensis, Enterobius vermicularis, Trichuris suis and other species develop to the infective stage within the egg. The infective juvenile hatches when it receives the appropriate stimulus from the host. The whole of the free-living phase of the life cycle is thus protected by a resistant egg-shell. The egg-shells of these species enable them to survive desiccation, freezing and mechanical and chemical damage (Wharton, 1980; Perry and Wharton, 1985).

A few nematodes with direct life cycles have eliminated the free-living phase altogether. Juveniles produced by the ovoviviparous female of *Trichinella spiralis* burrow through the walls of the intestine, migrate and encyst in the muscles of the host where they develop to the infective stage. The infection is acquired when the infected flesh is eaten by a carnivore. The infective juveniles of *Toxocara canis* can migrate across the placenta of the dog and infect pups *in utero.* There is also evidence that several nematode infections can be transmitted via the milk (Miller, 1981).

Indirect Life Cycles

Indirect life cycles involve one or more intermediate hosts, in addition to the definitive host in which sexual reproduction occurs. The inclusion of an intermediate host may remove the nematode from the hazards of the external environment for the greater part of its life cycle. The intermediate host is often the prey of the definitive host. The chances of transmission are increased by exploiting the natural food chain. Development may occur within the intermediate host but there is no increase in numbers. Digenean parasites, in contrast, often undergo substantial asexual reproduction within their intermediate hosts.

The eggs of *Metastrongylus apri*, the lungworm of pigs, hatch when ingested by an earthworm. The juveniles then moult twice to the infective 3rd stage. Pigs are infected if they ingest infected earthworms or the 3rd-stage juvenile, which can survive in the soil if the earthworm dies. The definitive hosts of *Anisakis marina* are fish-eating birds and mammals. The eggs hatch as 2nd-stage juveniles which are ingested by copepods and other marine invertebrates. If the copepod is ingested by a fish, the juvenile encysts within the tissues and the definitive host then acquires the infection if it eats an infected fish.

Vectors

Free-living stages have been completely eliminated from the life cycle of most filarid nematodes by the use of a blood-sucking insect as an intermediate host. The insect acts as a vector, transferring the infection from host to host when it takes a blood meal. The nematode is protected from the external environment for the whole of its life cycle, apart from a brief period when infective juveniles are deposited on the skin of the host and before they penetrate the wound made by the vector's feeding activities. The use of a vector greatly increases the chances of transmission, as the vector itself has sensory and behavioural mechanisms that enable it to locate the vertebrate host before feeding. Filarids cause some of the most important nematode diseases of humans (Table 5.4).

The adults of *Onchocerca volvulus*, the filarid causing river blindness in humans, live in nodules in the subcutaneous tissues. Females are ovoviviparous, producing 1st-stage juveniles called microfilariae. Microfilariae may retain the egg-shell as a sheath (are 'ensheathed'); the microfilariae of *O. volvulus* are unsheathed. The microfilariae are carried to the peripheral circulation where they may be ingested by the vector, *Simulium* spp. (blackflies). The nematode undergoes two moults within the vector to a 3rd-stage juvenile which migrates to the head and infects the labium of the insect's mouthparts. When the vector takes a blood meal, the infective juveniles pass down the labium and enter the body via the feeding wound. The final two moults occur within the human and adult worms appear in nodules within a year.

There is a distinct periodicity in the appearance of some filarid microfilariae in the peripheral circulation. *Wuchereria bancrofti* has a nocturnal periodicity with microfilariae appearing at night (Figure 6.9). *Loa loa* exhibits a diurnal periodicity. Microfiliarial periodicity increases the chances of transmission by ensuring that the greatest numbers are present in the peripheral circulation at the time when the vector is feeding. *Culex fatigens* and other mosquito vectors of *W. bancrofti* feed at night whereas *Chrysops* spp., the vector of *L. loa*, feed during the day.

When they are absent from the peripheral circulation, the

Figure 6.9: The Numbers of *Wuchereria bancrofti* Microfilariae in the Peripheral Circulation at Different Times of the Day. The microfilarial count exhibits a distinct periodicity, the peak of which coincides with the feeding activities of the vector. Redrawn from Cheng (1973)

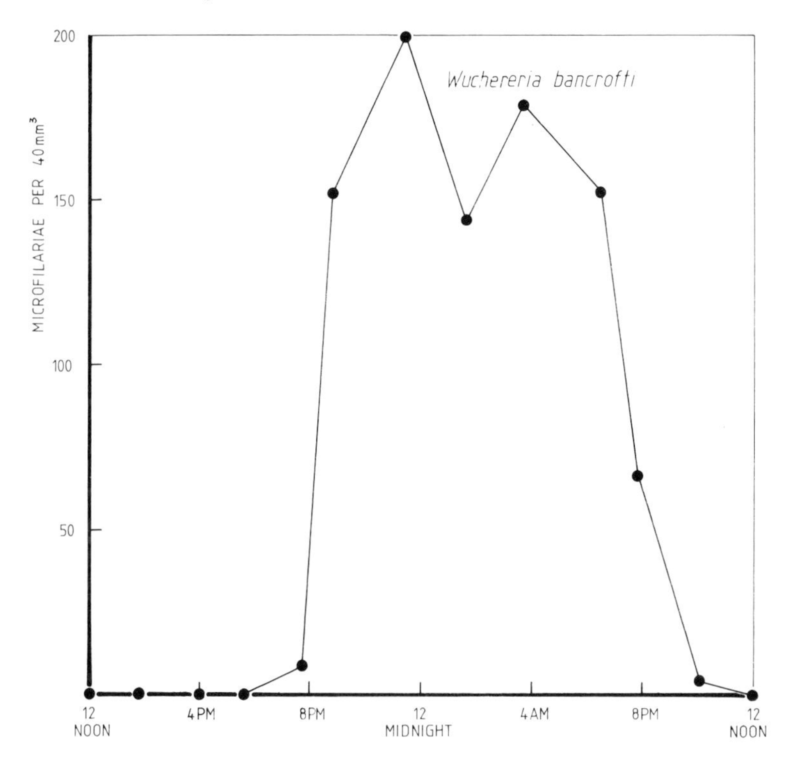

microfilariae of *W. bancrofti* can be found in the blood vessels around the lungs. There is evidence that the periodicity is a response to variations in oxygen tension within the lungs (Hawking, 1975). If an infected person breathes oxygen at night when microfilariae are present in the peripheral circulation, the peripheral microfilarial count rapidly falls. When oxygen breathing ceases, the count rises again.

These observations may be explained by oxygen tensions affecting the behaviour of the microfilariae. During the day, when oxygen tensions are high, the lungs present what Hawking (1975) calls an 'oxygen barrier'. This inhibits the passage of microfilariae through the lungs by causing a reversal of wave contraction. The microfilariae therefore tend to accumulate in the blood vessels around the lungs. At night oxygen tensions are lower (40-45 mmHg rather than 55 mmHg) and the microfilariae pass through the lungs more easily. They therefore become more evenly distributed throughout the vascular system, with a consequent rise in numbers in the peripheral circulation.

In species with different patterns of periodicity other mechanisms must be operating. Changes in body temperature may affect the response of microfilariae to the 'oxygen barrier'. There is also some indication that microfilariae possess an endogenous rhythm (Hawking, 1975).

7 ENVIRONMENTAL PHYSIOLOGY

Environmental conditions limit the survival, reproduction and growth of nematodes; as they do with all organisms. Hazards include temperature extremes, desiccation, flooding, osmotic and ionic stress, toxic chemicals, pathogens and predation. The distribution of free-living nematodes is limited by their ability to withstand such conditions, and parasitic species often have free-living stages which must be able to withstand conditions outside the host.

Different species of nematodes and different stages in the life cycle vary in their ability to survive environmental stresses and in the mechanisms by which they do so. Survival abilities have been classified in various ways. Terms have been defined by the level of metabolic activity present in an organism and by its effect on the ageing process (Keilin, 1959; Cooper and Van Gundy, 1971). An animal not under environmental stress will metabolise normally and age continuously throughout its lifespan. Under adverse environmental conditions the organism becomes quiescent with a reduced level of metabolic activity and a corresponding reduction in the rate of ageing. Under more severe environmental stress the organism may enter into a state of anabiosis or cryptobiosis. In the anabiotic state there is no detectable metabolism and the ageing process is suspended (Figure 7.1). An organism can survive in an anabiotic state for many years. The plant parasitic nematode, *Ditylenchus dipsaci*, has been stored dry for 23 years and yet will resume activity after 2 or 3 hours' immersion in water (Fielding, 1951).

There is a spectrum of metabolic activity and it may be difficult to distinguish between normal, quiescent and anabiotic states. Evans and Perry (1976) consider that a classification of dormant states based upon metabolic levels is unsatisfactory. Their classification extends that proposed by Laudien (1973), which categorises dormant states according to the nature of the physiological mechanism that initiates the developmental arrest. A quiescent state is a temporary interruption or slowing down of development as a result of unfavourable environmental conditions. Normal development resumes as soon as favourable conditions

Figure 7.1: Comparison of Normal, Quiescent and Anabiotic States. During quiescence metabolism is reduced and ageing slowed down. In an anabiotic state metabolism and ageing cease. Redrawn from Cooper and Van Gundy (1971)

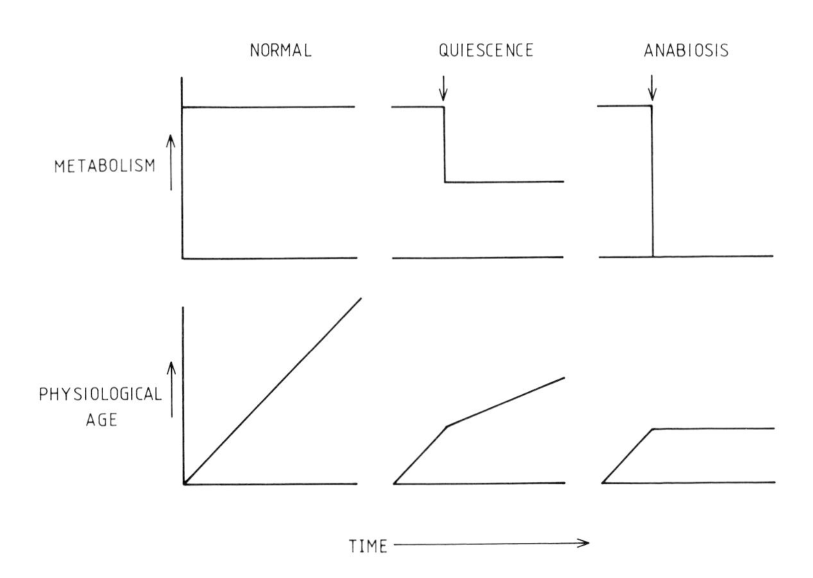

return. Diapause, however, is a developmental arrest which, although it may be initiated by environmental conditions, is not ended by a return to normal conditions but needs a specific environmental or physiological cue for development to resume.

Failure to distinguish between developmental events in the life cycle and the direct effect of the environment on metabolism has led to confusion in the use of these terms. A reduction in the level of metabolism will affect the rate of development and a developmental arrest can affect the level of metabolism but the two processes are fundamentally different.

A developmental arrest is an obligatory part of the life cycle of many nematodes. Development resumes only after a specific stimulus is received. The eggs of *Globodera rostochiensis* develop to the 2nd-stage juvenile and lie dormant within the cyst formed by the body of the female until they are stimulated to hatch by a factor present in the root diffusate of their potato host. The infective juveniles of trichostrongyle nematodes exsheath and develop to the 4th-stage juvenile in response to carbon dioxide within the gut of the host. The eggs of some plant parasitic nematodes require chilling before they will hatch. These all represent developmental arrests which are an obligatory part of the life cycle. Facultative developmental arrest is triggered by environmental cues. This induces diapause states such as the hatching of *Nematodirus* eggs and arrested development in trichostrongyle 4th-stage juveniles.

Although the classification of dormant states on the basis of metabolic levels (Keilin, 1959; Cooper and Van Gundy, 1971) is valid, confusion arises if states of developmental arrest are included in this scheme. It is more satisfactory to recognise two distinct types of dormant state: metabolic dormancy and developmental dormancy (Table 7.1). A nematode undergoing a developmental dormancy may at the same time be capable of periods of metabolic dormancy. Trichostrongyle infective juveniles are in a state of developmental dormancy, awaiting a stimulus from the host before completing the J2-J3 moult and commencing development to the 4th-stage juvenile. They are also capable of periods of metabolic dormancy and can survive extreme desiccation in a state of anabiosis (Wharton, 1982c).

Many animals, including mammals, can survive with a reduced

Table 7 1: States of Dormancy in Nematodes

Dormant state	Definition	Example
Metabolic dormancy		
Quiescence	Reduced metabolism in response to environmental conditions	Most species, e.g. in response to low temperatures
Anabiosis	Cessation of metabolism in response to extreme environmental conditions	Desiccation survival in *D. dipsaci*
Developmental dormancy		
Obligative	A delay in development which is a normal part of the life cycle and is ended by a specific physiological trigger	Hatching of *G. rostochiensis*; exsheathment of trichostrongyle infective juveniles
Facultative (diapause)	A delay in development initiated by environmental conditions and ended spontaneously or by a specific physiological or environmental cue	Hatching of *N. battus* eggs; arrested development in trichostrongyle nematodes

level of metabolism (quiescence) but only a few groups of small invertebrates can survive in a state in which metabolism ceases altogether or is reduced to undetectable levels (anabiosis). Evans and Perry (1976) have argued that anabiosis is an extreme form of quiescence but the adaptations involved in surviving with a reduced level of metabolism may be quite different from those involved in surviving an interruption of metabolic activity. Separate terms should, therefore, be used for these phenomena.

Various types of anabiosis have been recognised, depending on the nature of the environmental stress responsible for the interruption of metabolism (Keilin, 1959). Anhydrobiosis, cryobiosis, osmobiosis and anoxybiosis result from desiccation, low temperature, osmotic stress and lack of oxygen, respectively. The mechanisms of anhydrobiosis have been the most thoroughly investigated.

Desiccation Survival and Anhydrobiosis

Field studies of nematode survival on pasture indicate that the free-living stages of many nematode parasites can survive periods of desiccation. A number of laboratory studies have attempted to define the survival abilities of different species and stages. These studies have used different relative humidities, temperatures and lengths of exposure, and it is, therefore, difficult to compare results. Comparison is aided by calculating the 50 per cent survival time (S_{50}) using the methods of probit analysis (Table 7.2). The free-living stages of several animal and plant parasitic nematodes can survive exposure to desiccation for long periods of time. Different stages in the life cycle differ in their ability to survive desiccation, and survival may be limited to a particular stage.

The 1st-stage juvenile within the egg and the infective 3rd-stage juvenile are the desiccation-tolerant stages of trichostrongyle nematodes (Wharton, 1982c). The hatched 1st-stage and the 2nd-stage juvenile are susceptible to desiccation. The infective stage can survive such extreme desiccation that it must be capable of anhydrobiosis. *Trichostrongylus colubriformis* infective juveniles can survive vacuum desiccation at 0 per cent relative humidity and 1.5 torr with an S_{50} of 8.8h. A number of plant parasitic nematodes are capable of anhydrobiosis and can survive in a desiccated state for many years (Table 7.3).

Table 7 2: Desiccation Survival in Nematodes

Species	Stage	Temperature (°C)	% Relative Humidity	S_{50} (Days)	Reference
Animal parasitic nematodes					
Trichostrongylus retortaeformis	J3	20-26	44	18.1	Wharton, 1982c
T. colubriformis	J3	20	33	84	Wharton, 1982c
	J2	20	98	<1 h	
	J1 in egg	20	76	5.8	
Ostertagia ostertagi	J3	27	50	4.9	Wharton, 1982c
O. circumcincta	J3	27	26	15.3	Wharton, 1982c
Haemonchus contortus	J3	18	47	<2	Wharton, 1982c
Nematodirus battus	J3 in egg	15	33	>105	Parkin, 1976
	Hatched J3	15	33	84	Wharton, 1982c
Plant parasitic nematodes					
Ditylenchus dipsaci	J4	18	45	12	Perry, 1977a
D. myceliophagus	J4	18	45	19 min	Perry, 1977a
Heterodera avenae	J2 in cyst	15	40	4.5years	Meagher, 1974

*For further references see Wharton (1982c)

Less is known about the survival abilities of free-living nematodes. Field studies indicate that desiccation survival is widespread among free-living nematodes and that a number of species can survive anhydrobiotically. The fungivorous nematode, *Aphelenchus avenae*, can survive desiccation if it is allowed to dry slowly (Crowe and Madin, 1975). Desiccation survival has been reported in the dauer larvae of rhabditid nematodes (Evans and Womersley, 1980) and various nematode species have been extracted from dry moss and desert soils. *Actinolaimus hintoni* and *Dorylaimus keilini* are freshwater nematodes which can survive several months of dry conditions in the temporary ponds found in Nigeria (Lee, 1961).

A slow rate of water loss is essential for nematodes to survive desiccation. If they are dried quickly, *A. avenae* die, but they enter a state of anhydrobiosis if allowed to dry at 98 per cent relative humidity in pellets before exposure to drier conditions (Crowe and Madin, 1975). *Panagrellus silusae* will only survive desiccation if dried slowly in agar (Lees, 1953). These species can only survive desiccation if environmental conditions are such that they lose water slowly, but some species can survive direct exposure to low

Table 7 3: Longevity of Plant Parasitic Nematodes in a State of Anhydrobiosis

Nematode	Longevity
Anguina agrostis	4 years
A. amsinckia	4.3 years
A. balsamophila	24 years
A. tritici	32 years
Aphelenchoides subtenuis	3 years
A. xylophilus	1-2 years
Circonemoides xenoplox	2 years
Ditylenchus dipsaci	23 years
D. triformis	2.5 years
Helicotylenchus dihystera	8 months
Hemicycliophora avenaria	6 months
Heterodera avenae	5.5 years
H. qlycines	6 years
H. schachtii	3 months
Globodera rostochiensis	2.5 years
Pratylenchus dianthus	4.5 years
P. projectus	3.5 months
P. penetrans	11 months
P. thornei	1 year
Radopholus similis	6 months
Sphaeronema californicum	1 year
Tylenchus polyhypnus	5 months
Tylenchus polyhypnus	39 years
Xiphenema americanum	9 months
X. bakeri	8 months
X. index	6-9 months

Data from Norton (1978)

relative humidities. The infective juveniles of *D. dipsaci* lose water relatively slowly when exposed to 33 per cent relative humidity and will survive for several days (Perry, 1977a,b). This species has an intrinsic ability to control the rate of water loss.

Species which are susceptible to rapid desiccation may be able to survive under conditions of slow water loss but in most species this has not been tested. Once *A. avenae* has entered an anhydrobiotic state as a result of slow water loss, it will survive exposure to severe desiccation (Crowe and Madin, 1975). The ability to survive desiccation and to enter a state of anhydrobiosis must be much more widespread than has been realised.

Mechanisms of Anhydrobiotic Survival

The survival abilities of an organism in a state of anhydrobiosis are truly remarkable. For example, *D. dipsaci* can survive with a water content of less than 1 per cent and with no detectable metabolism. The sensitivity of techniques used to detect metabolism suggests that metabolic levels are, at best, less than 0.01 per cent of normal. Metabolism starts immediately upon immersion in water, however, and activity resumes after a lag phase of 2 to 3 h (Barrett, 1982). We have little idea how an organism can survive in this remarkable way, although it is generally agreed that a slow rate of water loss is important. This may be controlled by the animal itself or it may be a result of environmental conditions.

Some nematodes form tight coils when exposed to desiccation. This is thought to slow down the rate of water loss by reducing the area of cuticle directly exposed to air (Wharton, 1982a). Nematodes that can control the rate of water loss appear to have a restricted cuticular permeability. Ellenby (1968) has suggested that the permeability of the cuticle of *D. dipsaci* decreases as it dries. The epicuticle of *Anguina tritici* is much more prominent in anhydrobiotic than in hydrated specimens (Bird and Buttrose, 1974), which may indicate a change in cuticular permeability.

Aphelenchus avenae must be dried slowly over a 72h period before they can survive anhydrobiotically. Glycogen and lipid levels decrease during this period whereas those of trehalose and glycerol increase markedly (Figure 7.2). These changes are reversed when nematodes are reimmersed in water (Madin, Crowe and Loomis, 1978). It has been suggested that glycerol replaces 'bound water' which is intimately associated with biological macromolecules (Womersley, 1981). Glycerol levels are, however, low in some anhydrobiotic nematodes and in these species inositol or another polyhydric alcohol may be acting as a replacement for bound water (Womersley, 1981). Although *P. silusae* has relatively high levels of glycerol, it is susceptible to desiccation (Barrett, 1982). Glycerol appears to be produced by nematodes in response to a variety of environmental stresses (J. Barrett, pers. comm.) and is not solely involved in desiccation survival.

Trehalose may stabilise membranes in the dry state (Crowe, Crowe and Chapman, 1984). It has been suggested that hydrogen bonding between trehalose and the phosphate head groups of phospholipids replaces that between the lipid and water. Trehalose is, however, widely distributed among nematodes, acanthocephalans and arthropods. For example, it makes up 70-80 per cent of the total carbohydrate content of the testis and seminal

Figure 7.2: Changes in the Concentrations of Various Metabolites during Exposure of *A. avenae* to Desiccating Conditions. Lipid and glycogen levels decrease whereas those of glycerol and trehalose increase. Redrawn from Madin and Crowe (1975)

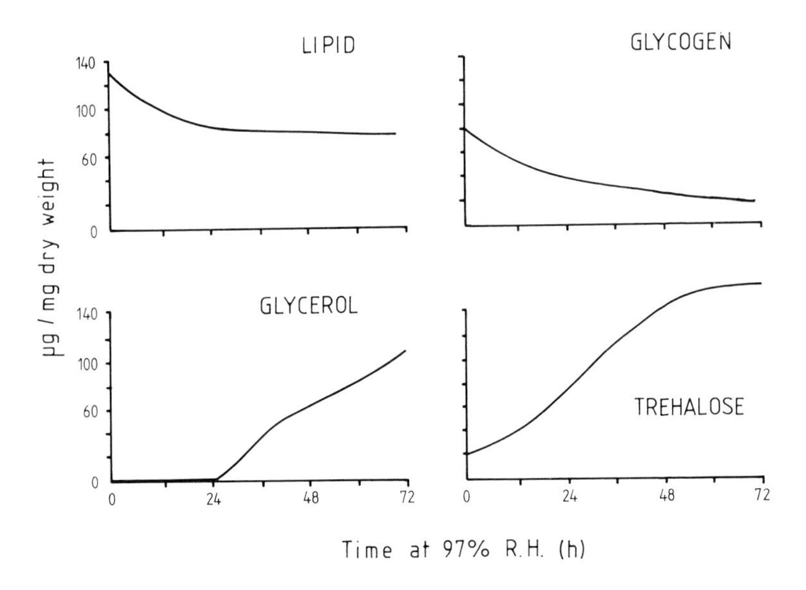

vesicle of *Ascaris* (Barrett, 1981). Trehalose is found in species that have little ability to survive desiccation.

The accumulation of trehalose and glycerol may be a purely fortuitous result of metabolic disruption during entry into anhydrobiosis. It may, therefore, be unwise to designate a specific protective compound that enables an animal to survive anhydrobiotically. A whole complex of morphological, physiological and biochemical mechanisms may be involved. In *D. dipsaci*, enzymes are not denatured by desiccation, and comparison of metabolite profiles in the dry and hydrated state does not indicate specific biochemical control during anhydrobiosis (Barrett, 1982).

Ultrastructural observations show that tissues condense and pack together in the anhydrobiotic state (Bird and Buttrose, 1974). During the lag phase that occurs before the recovery of *D. dipsaci* after immersion in water, there is a marked change in the appearance of the infective juvenile (Plate 7.1). A hyaline layer appears between the cuticle and the intestine (Wharton, Barrett and Perry, 1985). At the ultrastructural level, this can be seen to be a result of an increase in the thickness of the muscle cells with

Plate 7.1: Comparison of the Morphology of *Ditylenchus dipsaci* 4th-stage Juveniles. a, Undesiccated (isolated from fresh plant material and never exposed to desiccation); b, anabiotic (< 1% water content, mounted in glycerol); c, rehydrated (activity resumed after 3 h immersion of anabiotic juveniles in water). During rehydration a hyaline layer appears between the cuticle and intestinal cells (due to the expansion of the muscle cells and hypodermis) and a large, refringent granule appears within each of the intestinal cells (due to the coalescence of lipid droplets). × 350. From Wharton *et al.* (1985)

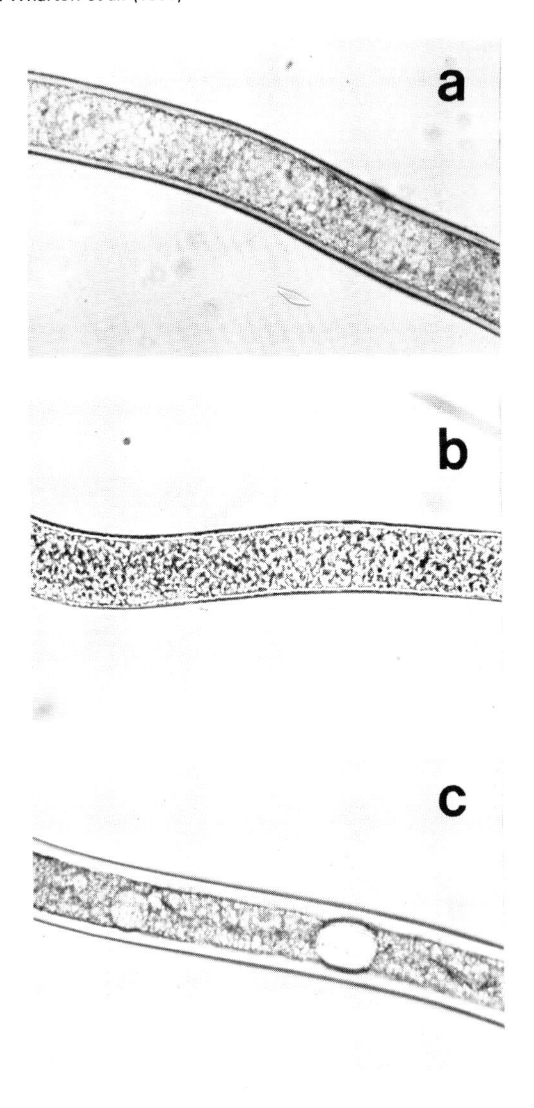

corresponding changes in muscle fibre spacing and in mitochondrial volume fraction and numerical density (Figure 7.3). These changes occur gradually throughout the lag phase and suggest that the muscle cells are repaired before movement begins (Wharton and Barrett, 1985). The nature of this repair is unknown but leakage of ions and metabolites during the early phase of rehydration and the re-establishment of the permeability barrier before movement recommences suggest that membrane repair is important. Recovery of normal muscle and nervous function is essential for movement to commence.

Cold Tolerance

Free-living nematodes and the free-living stages of parasitic nematodes will be exposed to sub-zero temperatures in temperate, arctic and sub-arctic regions of the world and at high altitudes. Although soil temperatures are significantly higher than air temperatures, particularly if the ground is covered by a layer of snow, nematodes may have to tolerate hard frost conditions; especially if they infect the aerial parts of plants or overwinter on pasture. The free-living stages of a number of animal and plant parasitic nematodes have been shown to survive sub-zero temperatures under laboratory conditions (Table 7.4). The infective juvenile appears to be the most resistant stage.

Arthropods that can survive sub-zero temperatures (cold tolerant) do so by withstanding extracellular ice formation (freezing tolerant) or by avoiding freezing at sub-zero temperatures by supercooling but die once the body freezes (freezing susceptible). An animal is said to supercool when it can maintain its body fluids in a liquid phase at temperatures below the melting point (Block, 1982). The temperature at which the body eventually freezes (the supercooling point) may be as low as −51°C in some insects.

Supercooling has only been investigated in a few species of nematode (Table 7.5) but these include free-living, and animal and plant parasitic species. These species will survive sub-zero temperatures down to the supercooling point but die once the body contents freeze. Nematodes are freezing susceptible but can avoid freezing by supercooling. It is expected that supercooling is a widespread method of cold tolerance in nematodes. *Aphelenchus ritzema-bosi* is the only species that has been suggested to be

Figure 7.3: Changes in the Dimensions of the Muscle Cells of *D. dipsaci* 4th-stage Juveniles during Rehydration and Recovery from Anabiosis. (a) Overall thickness (top line) and the thickness of the contractile zone increase (bottom line). (b) The spacing of the thick myofilaments increases (solid line) with a corresponding decrease in the muscle fibre numerical density (dotted line). U, Undesiccated (never exposed to desiccation); A, anabiotic (desiccated and prepared for electron microscopy using anhydrous techniques); R, rehydrated (active nematodes recovered from a state of anabiosis). Redrawn from Wharton and Barrett (1985)

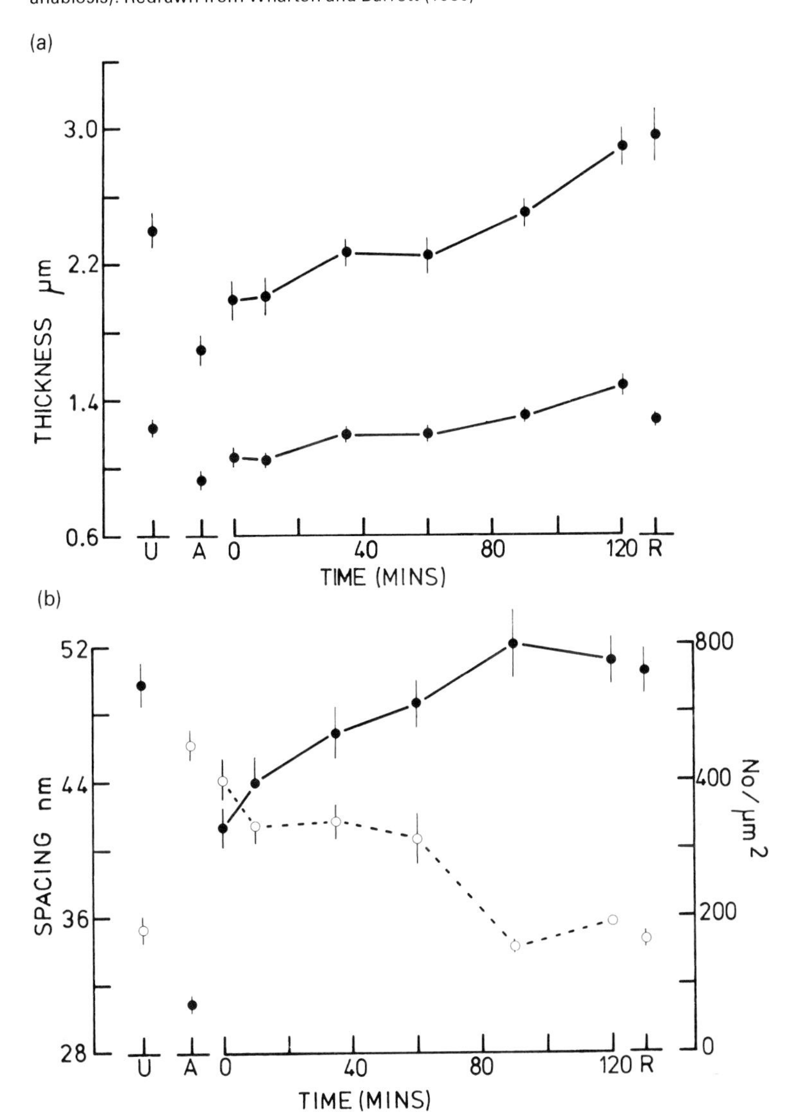

Table 7.4: Some Examples of Cold Tolerance in Nematodes

Species	Stage	Temperature (°C)	Time (days)	Reference
Animal parasitic nematodes				
Trichostrongylus colubriformis	J3	−10	16	Anderson, Wang and Levine, 1966
Nematodirus spathiger	J3 in egg	−10	14	Fallis, 1938
Anisakis sp.	Encysted J3	−10	12	Gustafson, 1953
Dochanoides stenocephala	J3	−20	30	Balasingham, 1964
Plant parasitic nematodes				
Meloidogyne hapla	J2 in egg	−4	10	Vrain, 1978
Bursaphelenchus lignicolis	J3	−10	5	Shôji, 1979
Heterodera tabacum	J2 in egg	−27	14	Miller, 1968

Table 7.5: Supercooling Points Recorded in Nematodes

Species	Stage	Supercooling point ± SEM	Reference
Trichostrongylus colubriformis	J3	−30.0 ± 0.7	Wharton *et al.*, 1984
Ditylenchus dipsaci	J4	−21.7 ± 0.7	Wharton *et al.*, 1984
Panagrellus silusae	Adult	−20.7 ± 0.84	Wharton *et al.*, 1984
Nematodirus battus	J3 in egg	−34.5 ± 0.49	Ash and Atkinson, 1982
Globodera rostochiensis	J2	−29.5 ± 1.03	Perry and Wharton, 1985

SEM, standard error of the mean

freezing tolerant (Asahina, 1959).

Most nematodes would be exposed to sub-zero temperatures in the presence of external water. External ice formation might initiate the freezing of the body contents by providing a nucleus, across the cuticle, for ice crystals to form in the supercooled fluid, in a similar fashion to a crystal seeding a supersaturated solution (inoculative freezing). The survival of sub-zero temperatures by *D. dipsaci* J4 and *T. colubriformis* J3 is similar whether they are in contact with water or not. These species are thus able to prevent inoculative freezing across the cuticle. However, *P. silusae* adults cannot survive if external ice forms, and cannot prevent inoculative freezing (Wharton, Young and Barrett, 1984). Unhatched

2nd-stage juveniles of *G. rostochiensis* can survive freezing in contact with water down to their supercooling point, whereas hatched juveniles cannot. The egg-shell prevents inoculative freezing in this species (Perry and Wharton, 1985).

The supercooling abilities of arthropods may be enhanced by the synthesis of cryoprotective compounds, such as trehalose, glycerol and other polyhydric alcohols. These compounds are also found in nematodes and there are some indications that nematodes acclimatise to low temperatures by synthesising cryoprotectants and thus lowering their supercooling point. Chilling the eggs of *N. battus* lowers their supercooling point (Ash and Atkinson, 1982) and results in an increase in trehalose concentration (Ash and Atkinson, 1983). The adults of *P. silusae* cultured at 22°C have a lower supercooling point after transfer to 5°C or 10°C (Mabbett and Wharton, unpublished observations).

The ability of nematodes to supercool and to prevent inoculative freezing must be important for them to survive over winter and this may explain important aspects of nematode biogeography. It would, for example, be interesting to compare the cold tolerance of antarctic, temperate and tropical species.

Osmotic and Ionic Regulation

Nematodes may have to tolerate considerable variations in the concentrations of salts and other chemicals in the environment and thus are subjected to a degree of osmotic and ionic stress. Infective stages, for example, are rapidly transferred from conditions in the soil or on pasture into those within the host. The infective juveniles of *T. colubriformis* rapidly lose water and the activity response to mechanical stimulation in hyperosmotic conditions (Figure 7.4). Loss of activity may be due to a decrease in volume, loss of turgor pressure, or the effect of increased ionic concentration on muscular or nervous activity (Wharton, Perry and Beane, 1983).

Osmoregulation

Studies on osmotic regulation in small nematodes are hampered by the difficulty of extracting body fluids for analysis. This has only been successfully achieved in the large ascarids, although it has been reported that fluid can be collected from nematodes with a volume as low as 0.1 μl (Wright and Newall, 1980). Most species

Figure 7.4: The Effect of Hyperosmotic Stress on the Activity Response to Mechanical Stimulus by the Infective Juveniles of *Trichostrongylus colubriformis*. (○) Distilled water, (●) artificial tap water, (■) 0.2MNaCl in artificial tap water, (□) 0.4MNaCl in artificial tap water. Redrawn from Wharton *et al.* (1983)

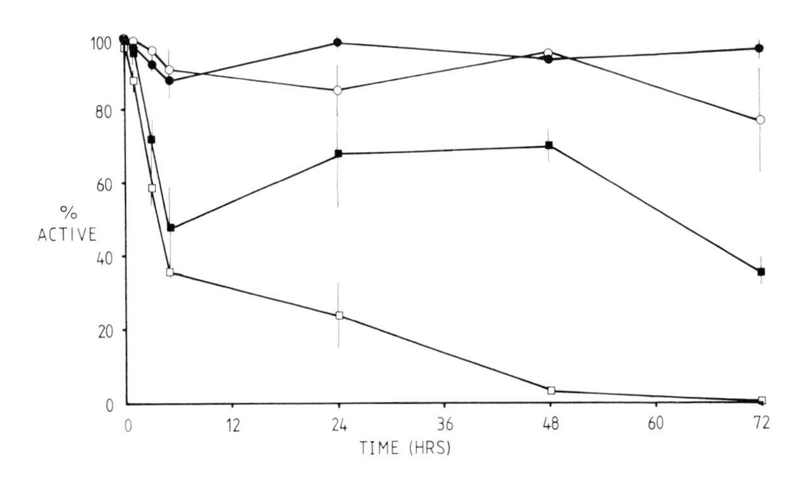

are even smaller than this. The high turgor pressure within the pseudocoel causes the rapid expulsion of body contents once the cuticle is punctured, again making fluid collection difficult.

Most studies have relied on measuring length or volume changes under hyperosmotic and hyposmotic conditions. An ability to osmoregulate is indicated by the maintenance of a constant volume in media of different osmotic concentrations. Volume changes may not, however, be a reliable measure of water flux. The limited compressibility of the cuticle may result in uneven collapse of the body as it loses water, making accurate volume determination difficult. Water flux can be measured directly using tritiated water or by the comparison of wet and dry weights (Wright and Newall, 1980). These methods require large numbers of nematodes and can take no account of variations between individuals. Interference microscopy allows the water content of individual nematodes to be measured and may be the best method of determining water flux (Wharton *et al.*, 1983).

The osmoregulatory abilities of various nematodes are shown in Table 7.6. Marine nematodes, in common with most marine invertebrates, are isosmotic with the environment (sea water). There is, therefore, no net osmotic gradient. Some marine nema-

Table 7.6: Osmoregulatory Abilities of Nematodes

Species	Environment	Relation of body fluids to environment	Measurement technique	Hyposmotic regulation	Hyperosmotic regulation	Reference
Deontostoma californicum	Marine	Isosmotic	V	–	++	Wright and Newall, 1976
D. timmerchioi	Marine	Isosmotic	V	–	++	Wright and Newall, 1976
D. antarcticum	Marine	Isosmotic	V	–	++	Wright and Newall, 1976
Monhystera disjuncta	Marine	Isosmotic	V	–	–	Wright and Newall, 1976
Enoplus communis	Marine	Isosmotic	V	–	+	Wright and Newall, 1976
E. brevis	30% sea water	Hyper-/isosmotic	V	++	+	Wright and Newall, 1976
Rhabditis terrestris	Soil	Hyperosmotic	V	++	+	Wright and Newall, 1976
Hammerschmidtiella diesingi	Cockroach hindgut	Isosmotic	V	–	+	Wright and Newall, 1976
Aspiculuris tetraptera	Mouse intestine	Isosmotic	V	++	++	Wright and Newall, 1976
Ascaris megalocephala	Horse intestine	Hyposmotic	V	–	–	Wright and Newall, 1976
Ascaris lumbricoides	Pig intestine	Hypo-/hyperosmotic	V	–	–	Wright and Newall, 1976
Nippostrongylus brasiliensis	Soil (J3)	Hyperosmotic	I	++	?	Atkinson and Onwuliri, 1981
Haemonchus contortus	Soil (J3)	Hyperosmotic	I	++	?	Atkinson and Onwuliri, 1981
Globodera rostochiensis	Soil (J2)	Hyperosmotic	I	++	+	Clarke, Perry and Hennessy, 1978
Heterodera schachtii	Soil (J2)	Hyperosmotic	I	++	+	Perry, Clarke and Hennessy, 1980
Trichostrongylus colubriformis	Soil (J3)	Hyperosmotic	I	++	–	Wharton *et al.*, 1983
Panagrellus silusae	Soil	Hyperosmotic	V	?	++	Prencepe, Bianco, Viglierchio and Scognamighio, 1984

–, No regulation; +, some regulation; ++, marked regulation; ?, not determined; V, volume/length measurements; I, interference microscope measurements.

todes have limited osmoregulatory abilities in different concentrations of sea water. *Enoplus brevis* is not exclusively marine, since it also occurs in salt marshes and estuaries where it has to face differing salinities. The ability of *E. brevis* to regulate its volume is much greater than the exclusively marine species, *E. communis* (Wright and Newall, 1976).

Soil nematodes are generally hyperosmotic to their environment. The concentration of solutes in soil water varies according to the degree of saturation of the soil. Rainfall will rapidly decrease osmotic concentrations, whereas evaporation will increase the concentration of solutes. The greatest osmotic problem, however, is the influx of water under hyposmotic conditions. Some species possess specific mechanisms for the removal of excess water.

The problem of water influx can be reduced by having a restricted cuticular permeability, which in some species is very limited. This appears to be related to the degree of osmotic stress and other environmental hazards (desiccation, toxic chemicals, etc.) which the nematode faces. The permeability constants of only a few species have been measured. The cuticles of marine nematodes have relatively high permeabilities with permeability constants of 3.4×10^{-4}cm/s for *E. brevis* and *E. communis.* Soil nematodes and the free-living stages of parasitic nematodes have much lower permeabilities. Permeability constants are 1.3×10^{-6}cm/s in *Aphelenchus avenae,* 2.0×10^{-6}cm/s in *Caenorhabditis elegans,* 6.1×10^{-6}cm/s in *Nippostrongylus muris* J3, and 0.8×10^{-6}cm/s in *Ancylostoma caninum* (Wright and Newall, 1980).

Some nematodes actively remove excess water under hyposmotic conditions. The intestine is a possible route for water removal, and rapid pumping of the intestine has been observed in *Rhabditis terrestris* in distilled water (Wright and Newall, 1980). The infective juveniles of trichostrongyle nematodes possess a tubular excretory system, the proximal part of which (the excretory ampulla) can be seen to pulsate. The pulsation rate is related to the degree of hyposmotic stress (Weinstein, 1952; Atkinson and Onwuliri, 1981) and it has been suggested that the excretory ampulla 'pumps out' excess water. Careful analysis of the pulsation cycle by video microscopy, however, has shown that pulsations are due to a passive cycle of filling and emptying of the ampulla (Figure 7.5). Water is passed into the ampulla continuously from the lateral excretory ducts; filling and emptying of the ampulla is

Figure 7.5: The Pulsation Cycle of the Excretory Ampulla of the Infective Juvenile of *Haemonchus contortus*. Drawings made from a video recording of the ampulla made using Nomarski interference contrast microscopy. (a) Excretory valve closed, excretory ampulla empty; (b) excretory ampulla dilates to its maximum volume; (c) excretory valve opens; (d) excretory ampulla empties; (a) excretory valve closes and cycle recommences. c, body cuticle; ea, excretory ampulla; ec, excretory cell; ep, excretory pore; ev, excretory valve

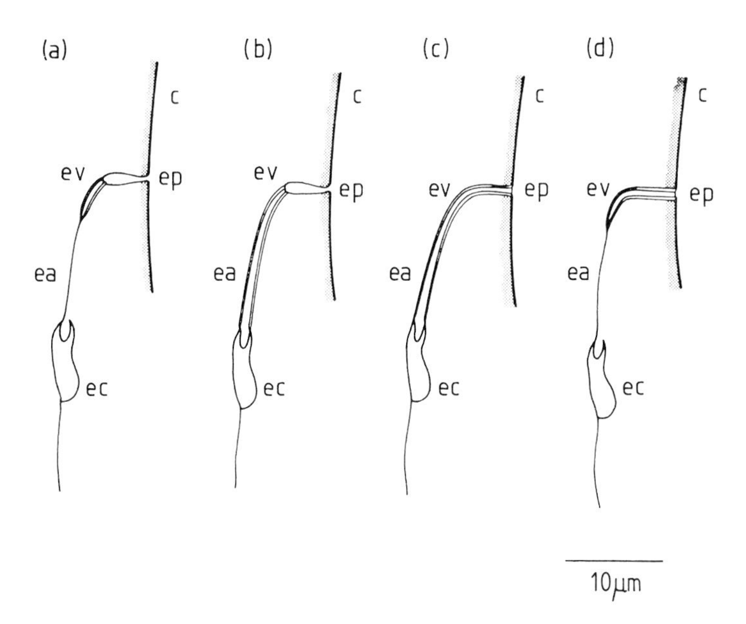

Reproduced from Wharton and Sommerville (1984), with permission

controlled by the opening and closing of the excretory valve. This system has structural and functional similarities with the contractile vacuole complex of protozoa (Wharton and Sommerville, 1984).

Influx of water under hyposmotic conditions may be reduced by excreting salts and thus decreasing the osmotic gradient (Oglesby, 1981). This is usually achieved by producing a urine but there is, as yet, little evidence for urine production in nematodes. Several possible sites have been suggested, including the intestine, hypodermis, the bacillary band cells of trichurids and the excretory system (Wright and Newall, 1980). There is, however, little physiological evidence of an excretory function for any of these structures.

Ionic Regulation

Lack of suitable methods has also hampered the study of ionic regulation in nematodes. Total ionic concentrations within a sample of nematodes can be determined spectrophotometrically or within a single nematode by X-ray microanalysis, but there is no method of determining ionic concentrations within the different body compartments of a small nematode. In the absence of suitable methods there is only circumstantial evidence for ionic regulation in most nematodes. X-ray microanalysis of 1 μm frozen-hydrated sections of quench-frozen specimens has proved a successful method for measuring element concentrations in different fluid compartments of a variety of tissues (Gupta and Hall, 1981). This, or a similar technique, may be the only feasible method of measuring elemental concentrations in different parts of a nematode. Wright and Newall (1980) have suggested the use of ion-specific microelectrodes but insertion of such electrodes, even with tip diameters as small as 1 μm, would probably be as difficult as removal of fluid for analysis.

Fluid extraction and analysis have been achieved in the large ascarids. *Ascaris* appears to be able to maintain constant levels of Na^+, Ca^{2+}, K^+ and Mg^{2+} within the pseudocoelomic fluid with different concentrations in the external medium (Hobson, Stephenson and Eden, 1952). In smaller nematodes it is only possible to measure the total concentration of salts. Changes in total Na^+ concentration indicate that *E. communis* conforms to changes in external concentration, whereas *E. brevis* shows some degree of regulation (Wright and Newall, 1980). Volume regulation only occurs in these species when Ca^{2+} and K^+ are present in the medium, in addition to Na^+. There is also some evidence for K^+ regulation in *P. redividus* and *A. avenae*, and for Na^+ regulation in *G. rostochiensis* (Wright and Newall, 1980).

Anaerobiosis

The mechanisms by which animal parasitic nematodes survive low oxygen tensions within their hosts have already been described (Chapter 5). Free-living nematodes are also found in environments where oxygen is in short supply. Oxygen penetration into marine and freshwater muds is limited, and nematodes have been found in anoxic conditions in such sites (Atkinson, 1976). The oxygen tension within soil depends on the degree of saturation with water and on the level of activity of micro-organisms.

Activity and development in nematodes require oxygen but some species can survive periods of anoxia in an inactive, anabiotic state. The motoneurons and muscles of nematodes are in a peripheral position (Figure 1.2). Oxygen does not have to penetrate as far to supply these structures as if they were situated centrally. This reduces the partial pressure of oxygen (pO_z) required to supply the nerves and muscles to 20 per cent of that required to supply all the body tissues (Atkinson, 1980). The oxygen tension required for activity may be further reduced by elaborations of the cuticle and by the long, thin body shape of nematodes, which ensure a high surface-to-volume ratio. Many nematodes are thus able to remain active under relatively low oxygen tensions. Growth and reproduction may be more susceptible to oxygen lack.

Animals may respond to lowered oxygen tensions by utilising oxygen at a rate proportional to the pO_z (respiratory conformers) or they maintain a constant consumption with falling pO_z (respiratory regulators). Regulation is achieved by the use of a respiratory pigment (haemoglobin) which binds and mediates the transport of oxygen, or by the presence of cytochromes which have a high affinity for oxygen and are still saturated at low oxygen tensions. The interpretation of respiratory responses to low oxygen tensions is confused by variations in the levels of activity exhibited by nematodes and the relatively low oxygen tensions required because of surface/volume effects. A considerable decrease in the pO_z may be possible before oxygen tensions in the tissues become limiting. Some species can supplement the oxygen supply by swallowing air. *Mesodiplogaster iheriteri* will swallow air at an air-agar interface, the rate of swallowing being influenced by oxygen availability (Klinger and Kunz, 1974).

Haemoglobin has been found in several animal parasitic nematodes but *E. brevis* is the only free-living nematode that has been shown to contain appreciable quantities of this pigment (Atkinson, 1976). It is found mainly in the larger nematodes and it may, therefore, facilitate oxygen transport in species with a relatively long diffusion path. The haemoglobin of *E. brevis* is concentrated around the oesophagus and ensures that sufficient oxygen is available to this organ during feeding in conditions of low oxygen tension.

Under anaerobic conditions, free-living invertebrates may

accumulate metabolic end-products which are metabolised when oxygen becomes available again. This results in a rise in oxygen consumption above normal levels as this 'oxygen debt' is repaid (Barrett, 1981). Some nematodes show such a rise in oxygen consumption after a period of anaerobiosis, and others do not. Increased activity will also result in a rise in oxygen consumption and this does not necessarily indicate an oxygen debt (Atkinson, 1980).

A few nematode species can survive long periods of anoxia in an anabiotic state. *Aphelenchus avenae* can survive anoxia for 30 days or longer (Cooper and Van Gundy, 1971). Little is known about the mechanisms of anoxybiotic survival. The fact that *A. avenae* switches from lipid to glycogen utilisation under anaerobic conditions may indicate a change from an aerobic respiratory metabolism to an anaerobic fermentative metabolism (Cooper and Van Gundy, 1971).

Pathogens and Predators

Interest in the predators and pathogens of nematodes has centred on their possible uses as biological control agents, particularly for the control of plant parasitic nematodes (Tribe, 1980). Fungal predators and pathogens are among the most promising from this point of view.

Nematophagous Fungi

Nematodes are attacked both by endoparasitic fungi, which infect the nematode via spores that adhere to the surface and fill the body with hyphae, and predatory fungi, which trap the nematode and absorb nutrients via invading hyphae. Nematode-trapping fungi use a variety of structures to capture nematodes, including erect branches, knobs, networks and constricting or non-constricting rings (Nicholas, 1984). Constricting rings consist of three cells which rapidly take up water, swell and trap the nematode when it enters the ring. Many traps produce adhesive substances that hold the nematode upon capture. Nordbring-Hertz and Mattiasson (1979) have suggested that lectins present on the traps bind to carbohydrate residues on the epicuticle of the nematode. They demonstrated that N-acetyl-galactosamine was bound most specifically by the trap lectins. N-acetyl-galactosamine residues have been demonstrated on the epicuticle of several plant parasitic nematodes (Spiegel, Cohn and Spiegel, 1982). The presence of carbohydrate-specific lectins on fungal traps suggests an interesting mechanism for host specificity.

Attempts to control plant parasitic nematodes using nematode-trapping fungi have largely been unsuccessful (Tribe, 1980). Some success has been achieved, however, in the commercial control of mycophagous nematodes on mushrooms and *Meloidogyne* spp. on horticultural crops. Nematophagous fungi act as a natural regulatory factor limiting nematode numbers in the soil, and may be responsible for recycling the nutrients held within the substantial biomass of nematodes (Mankau, 1980). Numbers of the cereal cyst nematode, *Heterodera avenae*, are suppressed under intensive cereal culture by the action of naturally occurring nematophagous fungi (Kerry, 1984).

The eggs of both animal and plant parasitic nematodes are parasitised by a variety of parasitic fungi (Lysek, 1978; Kerry, 1984). Eggs within the cyst of *G. rostochiensis* are often found to be parasitised by fungi, and, although they are yet to be cultured *in vitro*, these are promising biological control agents.

Other Pathogens

Nematodes are parasitised by viruses, rickettsias and bacteria, in addition to fungi (Mankau, 1980). *Bacillus penetrans* (now called *Pasturella penetrans*) is an obligate parasite which is specific to nematodes and is thus a promising biological control agent. There is some doubt concerning its taxonomic position but its ability to form vegetative microcolonies suggests that it has some affinities to the Actinomycetes. The nematode is infected by a spore which penetrates the cuticle by the growth of a spore germ tube. Microcolonies grow within the nematode and eventually fill the body with spores which are released when the body decomposes. Effective control of *Meloidogyne* spp. has been demonstrated in glasshouse tests (Mankau, 1980).

Predators

Predacious nematodes are found among the Mononchida and Dorylaimida. They feed on a variety of soil organisms, including other nematodes. The predacious mononchids are one of the commonest groups of soil nematodes and will attack most other groups of soil and plant parasitic nematodes. They have a large mouth armed with teeth which they use to catch and tear into their prey. The prey may be ingested whole or in pieces, or, in larger prey, the body wall is ruptured and the contents are sucked out. Predacious dorylaimids possess a stylet which is used to pierce the body wall of the prey. Secretions are injected which paralyse the prey and then the body contents are ingested via the stylet and the pumping of the oesophagus. Predacious nematodes are common in the soil and will readily attack the free-living stages of animal and plant parasitic nematodes. They are, however, non-specific and lack efficient prey location mechanisms. This limits their potential as a biological control agent but they certainly merit further investigation (Poinar, 1983).

A variety of other soil invertebrates are predators of nematodes, including mites, collembola, tardigrades, annelids, crustaceans and various insects. Some of these are very numerous and must be an important factor regulating nematode numbers in the soil. The difficulty of rearing many of these predators in the laboratory has prevented the investigation of their potential as biological control agents.

REFERENCES

Adams, J.M.P. and Tyler, S. (1980) 'Hopping Locomotion in a Nematode: Functional Anatomy of the Caudal Gland Apparatus of *Theristus caudacaliens* sp. n.', *Journal of Morphology, 164*, 265-85

Adamson, M.L. (1981) 'Studies on Gametogenesis in *Gyrinicola batrachiensis* (Walton, 1929) (Oxyuroidea: Nematoda)', *Canadian Journal of Zoology, 59*, 1368-76

Albertson, D.G. and Thomson, J.N. (1976) 'The Pharynx of *Caenorhabditis elegans*', *Philosophical Transactions of the Royal Society of London, B275*, 299-325

Anderson, F.L., Wang, G.T. and Levine, N.D. (1966) 'Effect of Temperature on Survival of the Free-living Stages of *Trichostrongylus colubriformis*', *Journal of Parasitology, 52*, 713-21

Anderson, R.C. (1984) 'The Origins of Zooparasitic Nematodes', *Canadian Journal of Zoology, 62*, 317-28

Anderson, R.C., Chabaud, A.G. and Willmott, S. (1974-83) *C.I.H. Keys to the Nematode Parasites of Vertebrates*, Nos. 1-10, Commonwealth Agricultural Bureaux, Farnham, UK

Anderson, R.M. (1978) 'The Regulation of Host Population Growth by Parasite Species', *Parasitology, 76*, 119-57

Anderson, R.M. (1982) 'The Population Dynamics and Control of Hookworm and Roundworm Infections', in R.M. Anderson (ed.), *The Population Dynamics of Infectious Diseases: Theory and Applications*, Chapman & Hall, New York and London

Anderson, R.M. and May, R.M. (1978) 'Regulation and Stability of Host-Parasite Population Interactions: I. Regulatory Processes', *Journal of Animal Ecology, 47*, 219-48

Anderson, R.V. and Coleman, D.C. (1981) 'Population Development and Interactions between Two Species of Bacteriophagic Nematodes', *Nematologica, 27*, 6-19

Anderson, R.V. and Darling, H.M. (1964) 'Embryology and Reproduction of *Ditylenchus destructor*, Thorne, with Emphasis on Gonad Development', *Proceedings of the Helminthological Society of Washington, 31*, 240-56

Andrassy, I. (1976) *Evolution as a Basis for the Systematisation of Nematodes*, Pitman, London

Anya, A.O. (1976) 'Physiological Aspects of Reproduction in Nematodes', *Advances in Parasitology, 14*, 267-351

Asahina, E. (1959) 'Frost-resistance in a Nematode, *Aphelenchus ritzema-bosi*', *Low Temperature Science, B17*, 51-62

Ash, C.P.J. and Atkinson, H.J. (1982) 'The Possible Role of Trehalose in the Survival of Eggs of *Nematodirus battus* during Dormancy', *Parasitology, 85*, lv

Ash, C.P.J. and Atkinson, H.J. (1983) 'Evidence for a Temperature-dependent Conversion of Lipid Reserves to Carbohydrate in Quiescent Eggs of the Nematode, *Nematodirus battus*', *Comparative Biochemistry and Physiology, 76B*, 603-10

Atkinson, H.J. (1976) 'The Respiratory Physiology of Nematodes', in N.A. Croll (ed.), *The Organisation of Nematodes*, Academic Press, New York and London

Atkinson, H.J. (1980) 'Respiration in Nematodes' in B.M. Zuckerman (ed.), *Nematodes as Biological Models*, Academic Press, New York and London

Atkinson, H.J. and Ballantyne, A.J. (1979) 'Evidence for the Involvement of Calcium in the Hatching of *Globodera rostochiensis*', *Annals of Applied Biology, 93*, 191-8
Atkinson, H.J. and Onwuliri, C.O.E. (1981) '*Nippostrongylus brasiliensis* and *Haemonchus contortus*: Function of the Excretory Ampulla of the Third-stage Larva', *Experimental Parasitology, 52*, 191-8
Atkinson, H.J. and Taylor, J.D. (1980) 'Evidence for a Calcium-binding Site on the Eggshell of *Globodera rostochiensis* with a Role in Hatching', *Annals of Applied Biology, 96*, 307-15
Awan, F.A. and Hominick, W.M. (1982) 'Observations on Tanning of the Potato Cyst-nematode, *Globodera rostochiensis*', *Parasitology, 85*, 67-71
Balasingham, E. (1964) 'Comparative Studies on the Effects of Temperature on Free-living Stages of *Plococonus laturis, Dochonoides stenocephela* and *Ancylostoma caninum*', *Canadian Journal of Zoology, 42*, 907-18
Barrett, J. (1976) 'Studies on the Induction of Permeability in *Ascaris lumbricoides* Eggs', *Parasitology, 73*, 109-21
Barrett, J. (1981) *Biochemistry of Parasitic Helminths*, Macmillan, London
Barrett, J. (1982) 'Metabolic Responses to Anabiosis in the Fourth Stage Juveniles of *Ditylenchus dipsaci* (Nematoda)', *Proceedings of the Royal Society of London, B216*, 159-77
Barrett, J. (1984) 'The Anaerobic End-products of Helminths', *Parasitology, 88*, 179-98
Barriga, O.O. (1981) *The Immunology of Parasitic Infections*, University Park Press, Baltimore
Barrington, E.J.W. (1979) *Invertebrate Structure and Function*, 2nd edn, Nelson, Sunbury-on-Thames
Befus, D. and Bienenstock, J. (1984) 'Induction and Expression of Mucosal Immune Responses and Inflammation to Parasitic Infections', in J.J. Marchalonis (ed.), *Contemporary Topics in Immunobiology*, Vol. 12, Plenum Press, New York and London
Behnke, J.M., Hannah, J. and Pritchard, D.I. (1983) '*Nematospiroides dubius* in the Mouse: Evidence that Adult Worms Depress the Expression of Homologous Immunity', *Parasite Immunology, 5*, 397-408
Bell, G. (1982) *The Masterpiece of Nature: the Evolution and Genetics of Sexuality*, Croom Helm, London
Bennet-Clark, H.C. (1976) 'Mechanics of Nematode Feeding', in N.A. Croll (ed.), *The Organisation of Nematodes*, Academic Press, New York and London
Bird, A.F. (1971) *The Structure of Nematodes*, Academic Press, New York and London
Bird, A.F. (1976) 'The Development and Organisation of Skeletal Structures in Nematodes', in N.A. Croll (ed.), *The Organisation of Nematodes*, Academic Press, New York and London
Bird, A.F. (1980) 'The Nematode Cuticle and its Surface', in B.M. Zuckerman (ed.), *Nematodes as Biological Models*, Vol. 2, Academic Press, New York and London
Bird, A.F. (1981) 'The *Anguina-Corynebacterium* Association', in B.M. Zuckerman and R.A. Rhode (eds), *Plant Parasitic Nematodes*, Vol. 3, Academic Press, New York and London
Bird, A.F. (1983) 'Growth and Moulting in Nematodes: Changes in the Dimensions and Morphology of *Rotylenchus reniformis* from Start to Finish of Moulting', *International Journal for Parasitology, 13*, 201-6
Bird, A.F. and Buttrose, M.S. (1974) 'Ultrastructural Changes in the Nematode *Anguina tritici* Associated with Anhydrobiosis', *Journal of Ultrastructural Research, 48*, 177-89

Bird, A.F. and McClure, M.A. (1976) 'The Tylenchid (Nematoda) Egg-shell: Structure, Composition and Permeability', *Parasitology, 72,* 19-28
Block, W. (1982) 'Cold Hardiness in Invertebrate Poikilotherms', *Comparative Biochemistry and Physiology, 73A,* 581-93
Brenner, S. (1974) 'The Genetics of *Caenorhabditis elegans*', *Genetics, 77,* 71-94
Byerley, L., Cassada, R.C. and Russell, R.C. (1976) 'The Life Cycle of the Nematode *Caenorhabditis elegans.* 1. Wild Type Growth and Reproduction', *Developmental Biology, 51,* 23-33
Calow, P. (1978) *Life Cycles,* Chapman & Hall, London
Calow, P. (1979) 'The Cost of Reproduction — a Physiological Approach', *Biological Reviews, 54,* 23-40
Calow, P. (1981) *Invertebrate Biology: a Functional Approach,* Croom Helm, London
Calow, P. (1983a) 'Energetics of Reproduction and its Evolutionary Implications', *Biological Journal of the Linnean Society, 20,* 153-65
Calow, P. (1983b) 'Pattern and Paradox in Parasite Reproduction', *Parasitology, 86,* 197-207
Cassada, R.C. and Russell, R.C. (1975) 'The Dauerlarva, a Post-embryonic Developmental Variant of the Nematode *Caenorhabditis elegans*', *Developmental Biology, 46,* 326-42
Castro, C.E. and Thomason, I.J. (1973) 'Permeation Dynamics and Osmoregulation in *Aphelenchus avenae*', *Nematologica, 19,* 100-8
Chapman, R.F. (1972) *The Insects: Structure and Function,* English Universities Press, London
Cheng, T.C. (1973) *General Parasitology,* Academic Press, New York and London
Chitwood, B.G. and Chitwood, M.B. (1974) *Introduction to Nematology,* University Park Press, Baltimore
Clark, R.B. (1967) *Dynamics in Metazoan Evolution: the Origin of the Coelom and Segments,* Clarendon Press, Oxford
Clark, W.C. (1978) 'Metabolite-mediated Density-dependent Sex Determination in a Free-living Nematode *Diplenteron potohikus*', *Journal of Zoology, 184,* 245-54
Clarke, A.J. and Perry, R.N. (1983) 'The Induction of Permeability in Egg-shells of *Ascaris suum*', *Parasitology, 87,* xxiv
Clarke, A.J. and Perry, R.N. (in press) 'Egg-shell Calcium and the Hatching of *Globodera rostochiensis*', *International Journal for Parasitology*
Clarke, A.J., Perry, R.N. and Hennessy, J. (1978) 'Osmotic Stress and the Hatching of *Globodera rostochiensis*', *Nematologica, 24,* 384-92
Cobb, N.A. (1915) 'Nematodes and their Relationships', *U.S. Department of Agriculture Yearbook,* 457-90
Conway Morris, S. (1981) 'Parasites and the Fossil Record', *Parasitology, 82,* 489-509
Cooper, A.F. and Van Gundy, S.D. (1971) 'Senescence, Quiescence and Cryptobiosis', in B.M. Zuckerman, W.F. Mai and R.A. Rhode (eds), *Plant Parasitic Nematodes,* Vol. 2, Academic Press, New York and London
Cowey, J.B. (1952) 'The Structure and Function of the Basement Membrane Muscle System in *Amphiporus lactifloreus* (Nemertea), *Quarterly Journal of Microscopical Science, 93,* 1-15
Crofton, H.D. (1966) *Nematodes,* Hutchinson, London
Crofton, H.D. (1971a) 'Form, Function and Behaviour', in B.M. Zuckerman, W.F. Mai and R.A. Rhode (eds), *Plant Parasitic Nematodes,* Vol. 1, Academic Press, New York and London
Crofton, H.D. (1971b) 'A Quantitative Approach to Parasitism', *Parasitology, 62,* 179-94

Croll, N.A. (1972) 'Behavioural Activities of Nematodes', *Helminthological Abstracts, A41*, 359-77

Croll, N.A. (1975a) Indolealkyamines in the Coordination of Nematode Behavioural Activities', *Canadian Journal of Zoology, 53*, 894-903

Croll, N.A. (1975b) Components and Patterns in the Behaviour of the Nematode *Caenorhabditis elegans*', *Journal of Zoology (London), 176*, 159-76

Croll, N.A. and Matthews, B.E. (1977) *Biology of Nematodes*, Blackie, Glasgow and London

Croll, N.A. and Smith, J.M. (1977) 'The Location of Parasites within their Hosts: the Behaviour of *Nippostrongylus brasiliensis* in the Anaesthetised Rat', *International Journal for Parasitology, 7*, 195-200

Croll, N.A. and Sukhdeo, V.K. (1981) 'Hierarchies in Nematode Behaviour', in B.M. Zuckerman and R.A. Rhode (eds), *Plant Parasitic Nematodes*, Vol. 3, Academic Press, New York and London

Crowe, J.H. and Madin, K.A.C. (1975) 'Anhydrobiosis in Nematodes: Evaporative Water Loss and Survival', *Journal of Experimental Zoology, 193*, 232-4

Crowe, J.H., Crowe, L.M. and Chapman, D. (1984) 'Preservation of Membranes in Anhydrobiotic Organisms: the Role of Trehalose', *Science, 223*, 701-3

Davey, K.G. (1982) 'Growth and Moulting in Nematodes', in E. Meerovitch (ed.), *Aspects of Parasitology*, McGill University, Montreal

Davey, K.G. and Rogers, W.P. (1982) 'Changes in Water Content Accompanying Exsheathment of *Haemonchus contortus*', *International Journal for Parasitology, 12*, 93-6

Davey, K.G. and Sommerville, R.I. (1982) 'Changes in Optical Path Difference in the Oesophageal Region and the Excretory Cells during Exsheathment in *Haemonchus contortus*', *International Journal for Parasitology, 12*, 503-7

del Castillo, J., de Mello, W.C. and Morales, T. (1964) 'Hyperpolarising Action Potentials Recorded from the Oesophagus of *Ascaris lumbricoides*', *Journal of General Physiology, 50*, 603-29

Dick, T.A. and Wright, K.A. (1974) 'The Ultrastructure of the Cuticle of the Nematode *Syphacia obvelata* (Rudolphi, 1802). IV. The Cuticle Associated with the Female Excretory Pore, Vulva and Vagina Vera', *Canadian Journal of Zoology, 52*, 245-50

Dusenbery, D.B. (1976) 'Attraction of the Nematode *Caenorhabditis elegans* to Pyridine', *Comparative Biochemistry and Physiology, 53C*, 1-2

Ehrenstein, von G. and Schierenberg, E. (1980) 'Cell Lineage and Development of *Caenorhabditis elegans*', in B.M. Zuckerman (ed.), *Nematodes as Biological Models*, Vol. 1, Academic Press, New York and London

Ellenby, C. (1968) 'Desiccation Survival in the Plant Parasitic Nematodes, *Heterodera rostochiensis* Wollenweber and *Ditylenchus dipsaci* (Kuhn) Filipjev' *Proceedings of the Royal Society of London, B169*, 203-13

Evans, A.A.F. and Perry, R.N. (1976) 'Survival Strategies in Nematodes', in N.A. Croll (ed.), *The Organisation of Nematodes*, Academic Press, New York and London

Evans, A.A.F. and Womersley, C. (1980) 'Longevity and Survival in Nematodes: Models and Mechanisms', in B.M. Zuckerman (ed.), *Nematodes as Biological Models*, Vol. 2, Academic Press, New York and London

Fallis, A.M. (1938) 'A Study on the Helminth Parasites of Lambs in Ontario', *Transactions of the Royal Canadian Institute, 22*, 81-128

Fielding, M.J. (1951) 'Observations on the Length of Dormancy in Certain Plant Infecting Nematodes', *Proceedings of the Helminthological Society of Washington, 18*, 110-12

Fleming, M.W. and Fetterer, R.H. (1984) '*Ascaris suum*: Continuous Perfusion of the Pseudocoel and Nutrient Absorption', *Experimental Parasitology, 57*, 142-8

Foor, W.E. (1968) 'Zygote Formation in *Ascaris lumbricoides*', *Journal of Cell Biology, 39*, 119-34

Foor, W.E. (1974) 'Morphological Changes of Spermatozoa in the Uterus and Glandular Vas Deferens of *Brugia pahangi*', *Journal of Parasitology, 60*, 125-33

Giebel, J. (1982) 'Mechanism of Resistance to Plant Nematodes', *Annual Review of Phytopathology, 20*, 257-79

Goldstein, P. (1981) 'Sex Determination in Nematodes', in B.M. Zuckerman and R.A. Rhode (eds), *Plant Parasitic Nematodes* Vol. 3, Academic Press, New York and London

Gray, J. (1968) *Animal Locomotion*, Weidenfeld and Nicolson, London

Green, C.D. (1980) 'Nematode Sex Attractants', *Helminthological Abstracts, B49*, 81-93

Gupta, B.J. and Hall, T.A. (1981) 'The X-ray Microanalysis of Frozen-hydrated Sections in Scanning Electron Microscopy: an Evaluation', *Tissue and Cell, 13*, 623-43

Gustafson, P.V. (1953) 'The Effect of Freezing on Encysted *Anisakis* Larvae', *Journal of Parasitology, 39*, 585-8

Harris, J.E. and Crofton, H.D. (1957) 'Internal Pressure and Cuticular Structure in *Ascaris*', *Journal of Experimental Biology, 34*, 116-30

Hawking, F. (1975) 'Circadian and Other Rhythms of Parasites', *Advances in Parasitology, 13*, 123-82

Heip, C., Smol, N. and Absillis, V. (1978) 'Influence of Temperature on the Reproductive Potential of *Oncholaimus oxyuris* (Nematoda: Oncholaimidae)', *Marine Biology, 45*, 255-60

Herman, R.K. and Horvitz, H.R. (1980) 'Genetic Analysis of *Caenorhabditis elegans*', in B.M. Zuckerman (ed.), *Nematodes as Biological Models*, Vol. 1, Academic Press, New York and London

Himmelhoch, S. and Zuckerman, B.M. (1983) '*Caenorhabditis elegans*: Characters of Negatively Charged Groups on the Cuticle and Intestine', *Experimental Parasitology, 55*, 299-305

Hinton, H.E. (1973) 'Neglected Phases in Metamorphosis: a Reply to V.B. Wigglesworth', *Journal of Entomology, A48*, 57-68

Hoagland, K.E. and Schad, G.A. (1978) '*Necator americanus* and *Ancylostoma duodenale*: Life History Parameters and Epidemiological Implications of Two Sympatric Hookworms of Humans', *Experimental Parasitology, 44*, 36-49

Hobson, A.D., Stephenson, W. and Eden, A. (1952) 'Studies on the Physiology of *Ascaris lumbricoides*. II. The Inorganic Composition of the Body Fluid in Relation to that of the Environment', *Journal of Experimental Biology, 29*, 22-9

Hope, W.D. (1974) 'Nematoda', in A.C. Giese and J.S. Pearse (eds), *Reproduction of Marine Invertebrates*, Vol. 1, Academic Press, New York and London

Howells, R.E. (1980) 'Filariae: Dynamics of the Surface', in H. van den Bossche (ed.), *The Host Invader Interplay*, Elsevier/North Holland, Amsterdam

Howells, R.E. and Blainey, L.S. (1983) 'The Moulting Process and the Phenomenon of Intermoult Growth in the Filarial Nematode *Brugia pahangi*', *Parasitology, 87*, 493-505

Huxley, H.E. (1953a) 'X-ray Analysis and the Problem of Muscle', *Proceedings of the Royal Society of London, B141*, 59-62

Huxley, H.E. (1953b) 'Electron Microscope Studies on the Organisation of Filaments in Striated Muscle', *Biochemical and Biophysical Acta, 12*, 387-94

Inglis, W.G. (1964) 'The Structure of the Nematode Cuticle', *Proceedings of the Zoological Society of London, 143*, 465-502

Inglis, W.G. (1983a) 'An Outline Classification of the Phylum Nematoda', *Australian Journal of Zoology, 31*, 243-55

Inglis, W.G. (1983b) 'The Design of the Nematode Body Wall: the Ontogeny of the Cuticle', *Australian Journal of Zoology, 31*, 705-16

Inglis, W.G. (1983c) 'The Structure and Operation of the Obliquely Striated Supercontractile Somatic Muscles in Nematodes', *Australian Journal of Zoology, 31*, 677-93

Inglis, W.G. (1985) 'Evolutionary Waves: Patterns in the Origins of Animal Phyla', *Australian Journal of Zoology, 33*, 153-78

Jarman, M. (1976) 'Neuromuscular Physiology of Nematodes', in N.A. Croll (ed.), *The Organisation of Nematodes*, Academic Press, New York and London

Jenkins, T. (1970) 'A Morphological and Histochemical Study of *Trichuris suis* (Schrank, 1788) with Special Reference to the Host-Parasite Relationship', *Parasitology, 61*, 357-74

Jenkins, T., Larkman, A. and Funnell, M. (1979) 'Spermatogenesis in a Trichuroid Nematode, *Trichuris muris*. 1. Fine Structure of Spermatogonia', *International Journal of Invertebrate Reproduction, 1*, 371-85

Johnson, C.D. and Stretton, A.O.W. (1980) 'Neural Control of Locomotion in *Ascaris*: Anatomy, Electrophysiology and Biochemistry', in B.M. Zuckerman (ed.), *Nematodes as Biological Models*, Vol. 1, Academic Press, New York and London

Jones, M.G.K. (1981) 'Host Cell Responses to Endoparasitic Nematode Attack: Structure and Function of Giant Cells and Syncytia', *Annals of Applied Biology, 97*, 353-72

Jones, M.G.K. and Dropkin, V.H. (1975) 'Cellular Alterations Induced in Soybean Roots by Three Endoparasitic Nematodes', *Physiological Plant Pathology, 5*, 119-24

Jones, M.G.K. and Dropkin, V.H. (1976) 'Scanning Electron Microscopy of Nematode-induced Giant Transfer Cells', *Cytobios, 15*, 149-61

Jungery, M., Mackenzie, C.D. and Ogilvie, B.M. (1983) 'Some Properties of the Surface of Nematode Larvae', *Journal of Helminthology, 57*, 291-5

Keilin, D. (1959) 'The Problem of Anabiosis or Latent Life: History and Current Concept', *Proceedings of the Royal Society of London, B150*, 149-91

Kennedy, C.R. (ed.) (1976) *Ecological Aspects of Parasitology*, North Holland Publishing Co., Amsterdam and Oxford

Kerry, B.R. (1984) 'Nematophagous Fungi and the Regulation of Nematode Populations in Soil', *Helminthological Abstracts, B53*, 1-14

Keymer, A. (1982) 'Density-dependent Mechanisms in the Regulation of Intestinal Helminth Populations', *Parasitology, 84*, 573-87

Kimble, J.E. and Hirsch, D. (1979) 'The Postembryonic Cell Lineages of the Hermaphrodite and Male Gonads of *Caenorhabditis elegans*', *Developmental Biology, 70*, 396-417

Kimble, J.E. and White, J.G. (1981) 'On the Control of Germ Cell Development in *Caenorhabditis elegans*', *Developmental Biology, 81*, 208-19

Klinger, J. and Kunz, P. (1974) 'Investigation with a Saprozoic Nematode *Mesodiplogaster iheriteri*, on a Possible Respiratory Function of Air Swallowing', *Nematologica, 20*, 52-60

Krupp, I.M. (1961) 'Effects of Crowding and of Superinfection on Habitat Selection and Egg Production in *Ancylostoma caninum*', *Journal of Parasitology, 47*, 957-61

Lackie, A.M. (1975) 'The Activation of Infective Stages of Endoparasites of Vertebrates', *Biological Reviews, 50*, 285-323

Laudien, H. (1973) 'Resting Stages in Development and their Induction or Termination by the Effect of Temperature', in H. Precht, J. Christopherson, H. Hensel and W. Larcher (eds), *Temperature and Life*, Springer-Verlag, Berlin and New York

Le Jambre, L.F. (1978) 'Anthelmintic Resistance in Gastrointestinal Nematodes of

Sheep', in A.D. Donald, W.H. Southcott and J.K. Dineen (eds), *The Epidemiology and Control of Gastrointestinal Parasites of Sheep in Australia*, CSIRO, Sydney
Lee, D.L. (1961) 'Two New Species of Cryptobiotic (Anabiotic) Freshwater Nematodes, *Actinolaimus hintoni* sp. nov. and *Dorylaimus keilini* sp. nov. (Dorylaimidae)', *Parasitology, 51*, 237-40
Lee, D.L. (1962) 'The Distribution of Esterase Enzymes in *Ascaris lumbricoides*', *Parasitology, 52*, 241-60
Lee, D.L. (1969) '*Nippostrongylus brasiliensis*: Some Aspects of the Fine Structure and Biology of the Infective Larva and the Adult', in A.E.R. Taylor (ed.), *Nippostrongylus and Toxoplasma*, Blackwell, Oxford
Lee, D.L. and Atkinson, H.J. (1976) *Physiology of Nematodes*, Macmillan, London
Lees, E. (1953) 'An Investigation into the Method of Dispersal and Desiccation Resistance of *Panagrellus silusae*', *Journal of Helminthology, 27*, 95-103
Levine, N.D. (1968) *Nematode Parasites of Domestic Animals and of Man*, Burgess, Minneapolis
Lewontin, R.C. (1965) 'Selection for Colonising Ability', in H.G. Baker and G.C. Stebbins (eds), *The Genetics of Colonising Species*, Academic Press, New York and London
Locke, M. (1982) 'Envelopes at Cell Surfaces — a Confused Area of Research of General Importance', in D.F. Metterick and S.S. Desser (eds), *Parasites: their World and Ours*, Elsevier Press, Amsterdam and New York
Lysek, H. (1978) 'A Scanning Electron Microscope Study of an Ovicidal Fungus on the Eggs of *Ascaris lumbricoides*', *Parasitology, 77*, 139-41
McClure, M.A. and Bird, A.F. (1976) 'The Tylenchid (Nematoda) Egg-shell: Formation of the Egg-shell in *Meloidogyne javanica*', *Parasitology, 72*, 29-39
McLaren, D.J. (1974) 'The Anterior Glands of Adult *Necator americanus* (Nematoda: Strongyloidea)', *International Journal for Parasitology, 4*, 25-37
McLaren, D.J. (1976) 'Sense Organs and their Secretions', in N.A. Croll (ed.), *The Organisation of Nematodes*, Academic Press, New York and London
McLaren, D.J. (1984) 'Disguise as an Evasive Stratagem of Parasitic Organisms', *Parasitology, 88*, 597-611
Madin, K.A.C. and Crowe, J.H. (1975) 'Anhydrobiosis in Nematodes: Carbohydrate and Lipid Metabolism during Dehydration', *Journal of Experimental Zoology, 193*, 335-42
Madin, K.A.C., Crowe, J.H. and Loomis, S.H. (1978) 'Metabolic Transition in a Nematode during Induction of and Recovery from Anhydrobiosis', in J.H. Crowe and J.S. Clegg (eds), *Dry Biological Systems*, Academic Press, New York and London
Maggenti, A.R. (1971) 'Nemic Relationships and the Origins of Plant Parasitic Nematodes', in B.M. Zuckerman and R.A. Rhode (eds), *Plant Parasitic Nematodes*, Vol. 1, Academic Press, New York and London
Maggenti, A.R. (1976) 'Taxonomic Position of Nematodes among the Pseudocoelomate Bilateria', in N.A. Croll (ed.), *The Organisation of Nematodes*, Academic Press, New York and London
Maggenti, A.R. (1981) *General Nematology*, Springer-Verlag, New York and Berlin
Mankau, R. (1980) 'Biological Control of Nematode Pests by Natural Enemies', *Annual Review of Phytopathology, 18*, 415-40
Mapes, C.J. (1965) 'Structure and Function in the Nematode Pharynx. 1. The Structure of the Pharynges of *Ascaris lumbricoides, Oxyuris equi, Apelectana brevicaudata* and *Panagrellus silusae*', *Parasitology, 55*, 269-84
Mapes, C.J. and Coop, R.L. (1973) 'On the Relationship Between Abomasal Electrolytes and Some Population Parameters of the Nematodes *Haemonchus*

contortus and *Nematodirus battus*', *Parasitology, 66*, 95-100

Marks, C.F., Thomason, I.J. and Castro, C.E. (1968) 'Dynamics of the Penetration of Nematodes by Water, Nematocides and Other Substances', *Experimental Parasitology, 22*, 321-37

Masamune, T., Anetai, M., Takasugi, M. and Katsui, N. (1982) 'Isolation of a Natural Hatching Stimulus, Glycinoeclepin A, for the Soybean Cyst Nematode', *Nature, 297*, 495-6

May, R.M. and Anderson, R.M. (1978) 'Regulation and Stability of Host-Parasite Population Interactions. II. Destabilizing Processes', *Journal of Animal Ecology, 47*, 249-68

Maynard Smith, J. (1978) *The Evolution of Sex*, Cambridge University Press, Cambridge

Meagher, J.W. (1974) 'Cryptobiosis of the Cereal Cyst Nematode (*Heterodera avenae*) and Effects of Temperature and Relative Humidity on Survival of Eggs in Storage', *Nematologica, 20*, 323-35

Meats, A. (1971) 'The Relative Importance to Population Increase of Fluctuations in Mortality, Fecundity and Time Variables of the Reproductive Schedule', *Oecologia (Berlin), 6*, 223-37

Mendis, A.H.W., Rose, M.E., Rees, H.H. and Goodwin, T.W. (1983) 'Ecdysteroids in Adults of the Nematode, *Dirofilaria immitis*', *Molecular and Biochemical Parasitology, 9*, 209-26

Michel, J.F. (1974) 'Arrested Development of Nematodes and Some Related Phenomena', *Advances in Parasitology, 12*, 279-366

Miller, G.C. (1981) 'Helminths and the Trans-mammary Route of Infection', *Parasitology, 82*, 335-42

Miller, P.M. (1968) 'The Susceptibility of Parasitic Nematodes to Sub-freezing Temperatures', *Plant Disease Reporter, 52*, 768-72

Moussa, M.T. (1969) 'Nematode Fossil Tracks of Eocene Age from Utah', *Nematologica, 15*, 376-80

Munn, E.A. (1977) 'A Helical, Polymeric Extracellular Protein Associated with the Luminal Surface of *Haemonchus contortus* Intestinal Cells', *Tissue and Cell, 9*, 23-34

Munn, E.A. and Greenwood, C.A. (1983) 'Endotube-Brush Border Complexes Dissected from the Intestines of *Haemonchus contortus* and *Ostertagia circumcincta*', *Parasitology, 87*, 129-37

Murrell, K.D. and Graham, C.E. (1983) 'Shedding of Antibody Complexes by *Strongyloides ratti* (Nematoda) Larvae', *Journal of Parasitology, 69*, 70-3

Neill, B.W. and Wright, K.A. (1973) 'Spermatogenesis in the Hologonic Testis of the Trichuroid Nematode, *Capillaria hepatica* (Bancroft; 1893)', *Journal of Ultrastructural Research, 44*, 210-34

Nelson, G.A., Roberts, T.M. and Ward, S. (1982) '*Caenorhabditis elegans* Spermatozoa Locomotion: Amoeboid Movement with Almost no Actin', *Journal of Cell Biology, 92*, 121-31

Nicholas, W.L. (1984) *The Biology of Free-living Nematodes*, Oxford University Press, Oxford

Nicholls, C.D., Lee, D.L. and Sharpe, M.J. (1985) 'Scanning Electron Microscopy of Biopsy Specimens Removed by a Colonoscope from the Abomasum of Sheep Infected with *Haemonchus contortus*', *Parasitology, 90*, 357-63

Nordbring-Hertz, B. and Mattiasson, B. (1979) 'Action of a Nematode-trapping Fungus Shows Lectin-mediated Host-Microorganism Interaction', *Nature, 281*, 477-9

Norton, D.C. (1978) *Ecology of Plant-parasitic Nematodes*, Wiley, New York

Ogilvie, B.M., Philipp, M., Jungery, M., Maizels, R.M., Worms, M.J. and Parkhouse, R.M.E. (1980) 'The Surface of Nematodes and the Immune Response of the

Host', in H. van den Bossche (ed.), *The Host Invader Interplay*, Elsevier/North Holland, Amsterdam

Oglesby, L.C. (1981) 'Volume Regulation in Aquatic Invertebrates', *Journal of Experimental Zoology, 215*, 289-301

O'Grady, R.T. (1983) 'Cuticular Changes and Structural Dynamics in the Fourth-stage Larvae and Adults of *Ascaris suum* Goeze, 1782 (Nematoda: Ascaroidea) Developing in Swine', *Canadian Journal of Zoology, 61*, 1293-1303

Osche, G. (1963) 'Morphological, Biological and Ecological Considerations in the Phylogeny of Parasitic Nematodes', in E.C. Dougherty (ed.), *The Lower Metazoa. Comparative Biology and Phylogeny*, University of California Press, Berkeley and Los Angeles

Parkin, J.T. (1976) 'The Effect of Moisture Supply upon the Development and Hatching of the Eggs of *Nematodirus battus*', *Parasitology, 73*, 343-54

Perry, R.N. (1977a) 'Desiccation Survival of Larval and Adult Stages of the Plant Parasitic Nematodes, *Ditylenchus dipsaci* and *D. myceliophagus*', *Parasitology, 74*, 139-48

Perry, R.N. (1977b) 'The Water Dynamics of Stages of *Ditylenchus dipsaci* and *D. myceliophagus* during Desiccation and Rehydration', *Parasitology, 75*, 45-70

Perry, R.N. and Clarke, A.J. (1981) 'Hatching Mechanisms of Nematodes', *Parasitology, 83*, 435-49

Perry, R.N. and Wharton, D.A. (1985) 'Cold Tolerance of Hatched and Unhatched Second Stage Juveniles of the Potato Cyst-nematode *Globodera rostochiensis*', *International Journal for Parasitology, 15*, 441-5

Perry, R.N., Clarke, A.J. and Hennessy, J. (1980) 'The Influence of Osmotic Pressure on the Hatching of *Heterodera schachtii*', *Revue de Nematologie, 3*, 3-9

Perry, R.N., Wharton, D.A. and Clarke, A.J. (1982) 'The Structure of the Egg-shell of *Globodera rostochiensis* (Nematoda: Tylenchida)', *International Journal for Parasitology, 12*, 481-5

Peters, W. (1978) 'Medical Aspects — Comments and Discussion II', in A.E.R. Taylor and R. Muller (eds), *The Relevance of Parasitology to Human Welfare Today*, Blackwell, Oxford

Philipp, M., Parkhouse, R.M.E. and Ogilvie, B.M. (1980) 'Changing Proteins on the Surface of a Parasitic Nematode', *Nature, 287*, 538-40

Poinar, G.O. (1979) *Nematodes for Biological Control of Insects*, CRC Press, Florida

Poinar, G.O. (1983) *The Natural History of Nematodes*, Prentice-Hall, New Jersey

Poinar, G.O. and Hansen, E. (1983) 'Sex and Reproductive Modifications in Nematodes', *Helminthological Abstracts, B52*, 145-63

Poole, R.W. (1974) *An Introduction to Quantitative Ecology*, McGraw-Hill, New York

Popham, J.D. and Webster, J.M. (1978) 'An Alternative Hypothesis of the Fine Structure of the Basal Zone of the Cuticle of the Dauer Larva of *Caenorhabditis elegans* (Nematoda)', *Canadian Journal of Zoology, 56*, 1556-63

Prencepe, A., Bianco, M., Viglierchio, D.R. and Scognamiglio, A. (1984) 'Response of the Nematode *Panagrellus silusae* to Hypotonic Solutions', *Proceedings of the Helminthological Society of Washington, 51*, 36-41

Preston, C.M. and Jenkins, T. (1983) 'Ultrastructural Studies of Early Stages of Oogenesis in a Trichuroid Nematode, *Trichuris muris*', *International Journal of Invertebrate Reproduction, 6*, 77-91

Prichard, R.K. (1978) 'Sheep Anthelmintics', in A.D. Donald, W.H. Southcott and J.K. Dineen (eds), *The Epidemiology and Control of Gastrointestinal Parasites of Sheep in Australia*, CSIRO, Sydney

Riddle, D.L. (1980) 'Developmental Genetics of *Caenorhabditis elegans*', in B.M. Zuckerman (ed.), *Nematodes as Biological Models*, Vol. 1, Academic Press, New

York and London
Riding, I. (1970) 'Microvilli on the Outside of a Nematode', *Nature, 226*, 179-80
Roberts, T.M. and Ward, S. (1982a) 'Membrane Flow during Nematode Spermiogenesis', *Journal of Cell Biology, 92*, 113-20
Roberts, T.M. and Ward, S. (1982b) 'Centripetal Flow of Pseudopodial Surface Components Could Propel the Amoeboid Movement of *Caenorhabditis elegans* Spermatozoa', *Journal of Cell Biology, 92*, 132-8
Rogers, W.P. (1960) 'The Physiology of the Infective Process of Nematode Parasites; the Stimulus from the Animal Host', *Proceedings of the Royal Society of London, B152*, 367-86
Rogers, W.P. (1982) 'Enzymes in the Exsheathing Fluid of Nematodes and their Biological Significance', *International Journal for Parasitology, 12*, 495-502
Rogers, W.P. and Petronijevic, T. (1982) 'The Infective Stage and the Development of Nematodes', in L.E.A. Symons, A.D. Donald and J.K. Dineen (eds), *Biology and Control of Endoparasites*, Academic Press, New York and London
Roggen, D.R. (1982) 'Functional Morphology of the Nematode Pharynx. 3. Which Model Fits?', *Nematologica, 28*, 326-32
Romeyn, K., Bouwman, L.A. and Admiraal, W. (1983) 'Ecology and Cultivation of the Herbivorous Brackish-water Nematode *Eudiplogaster pararmatus*', *Marine Ecology Progress Series, 12*, 145-53
Rosenbluth, J. (1965) 'Ultrastructure of the Somatic Muscle Cells in *Ascaris lumbricoides*. II. Intermuscular Junctions, Neuromuscular Junctions and Glycogen Stores', *Journal of Cell Biology, 26*, 579-91
Schad, G.A. (1977) 'The Role of Arrested Development in the Regulation of Nematode Populations', in G.W. Esch (ed.), *Regulation of Animal Populations*, Academic Press, New York and London
Schiemer, F. (1982a) 'Food Dependence and Energetics of Free-living nematodes. I. Respiration, Growth and Reproduction of *Caenorhabditis briggsae* (Nematoda) at Different Levels of Food Supply', *Oecologia (Berlin), 54*, 108-21
Schiemer, F. (1982b) 'Food Dependence and Energetics of Free-living Nematodes. II. Life History Parameters of *Caenorhabditis briggsae* (Nematoda) at Different Levels of Food Supply', *Oecologia (Berlin), 54*, 122-8
Schiemer, F. (1983) 'Comparative Aspects of Food Dependence and Energetics of Free-living Nematodes', *Oikos, 41*, 32-42
Schiemer, F., Duncan, A. and Klekowski, R.Z. (1980) 'A Bioenergetic Study of a Benthic Nematode, *Plectus palustris* de Mann 1880, throughout its Life Cycle. II. Growth, Fecundity and Energy Budgets at Different Densities of Bacterial Food and General Ecological Considerations', *Oecologia (Berlin), 44*, 205-12
Seymour, M.K. (1973) 'Motion and the Skeleton in Small Nematodes', *Nematologica, 19*, 43-8
Seymour, M.K. (1983) 'The Feeding Pump of *Ditylenchus dipsaci*', *Nematologica, 29*, 171-89
Seymour, M.K. and Shepherd, A.M. (1974) 'Cell Junctions Acting as Intestinal Valves in Nematodes', *Journal of Zoology (London), 173*, 517-23
Shelford, V.E. (1929) *Laboratory and Field Ecology*, Bailliere, Tindall and Cox, London
Shepherd, A.M., Clark, S.A. and Hooper, D.J. (1980) 'Structure of the Anterior Alimentary Tract of *Aphelenchoides blastophthorus* (Nematoda: Tylenchida, Aphelenchina)', *Nematologica, 26*, 313-57
Shôji, T. (1979) 'Resistance of Pine-wood Nematode, *Bursaphelenchus lignicolus*, to Low Temperatures', *Japanese Journal of Nematology, 9*, 5-8
Sibly, R. and Calow, P. (1985) 'Classification of Habitats by Selection Pressures: a Synthesis of Life-cycle and *r/K* Theory', in R.M. Sibly and R.H. Smith (eds), *Behavioural Ecology: Ecological Consequences of Adaptive Behaviour*,

Blackwell, Oxford
Siddiqui, M.R. (1983) 'Evolution of Plant Parasitism in Nematodes', in A.R. Stone, H.M. Platt and L.F. Khalil (eds), *Concepts in Nematode Systematics*, Academic Press, New York and London
Singh, R.N. and Sulston, J.E. (1978) 'Some Observations on Moulting in *Caenorhabditis elegans*', *Nematologica, 24*, 63-71
Smeal, M.G. and Donald, A.D. (1982) 'Inhibition of Development of *Ostertagia ostertagi* — Effect of Temperature on the Infective Larval Stage', *Parasitology, 85*, 27-32
Snell, T.W. (1978) 'Fecundity, Developmental Time, and Population Growth Rate', *Oecologia, 32*, 119-25
Sommerville, R.I. (1982) 'The Mechanics of Moulting in Nematodes', in E. Meerovitch (ed.), *Aspects of Parasitology*, McGill University, Montreal
Sommerville, R.I. and Weinstein, P.P. (1964) 'Reproductive Behaviour of *Nematospiroides dubius in vivo* and *in vitro*', *Journal of Parasitology, 50*, 401-9
Southey, J.F. (1978) *Plant Nematology*, HMSO, London
Spiegel, Y., Cohn, E. and Spiegel, S. (1982) 'Characterisation of Sialyl and Galactosyl Residues on the Body Wall of Different Plant Parasitic Nematodes', *Journal of Nematology, 14*, 33-9
Stretton, A.O.W., Fishpool, R.M., Southgate, E., Donmoyer, J.E., Walrond, J.P., Moses, J.E. and Kass, I.S. (1978) 'Structure and Physiological Activity of the Motoneurones of the Nematode *Ascaris*', *Proceedings of the National Academy of Science USA 75*, 3493-7
Sulston, J.E. and Horvitz, H.R. (1977) 'Post-embryonic Development of the Nematode *Caenorhabditis elegans*', *Developmental Biology, 56*, 110-56
Sulston, J.E. and White, J.G. (1980) 'Regulation and Cell Autonomy during Postembryonic Development of *Caenorhabditis elegans*', *Developmental Biology, 78*, 577-97
Sulston, J.E., Albertson, D.G. and Thomson, J.W. (1980) 'The *Caenorhabditis elegans* Male: Postembryonic Development of Non-Gonadal Structures', *Developmental Biology, 78*, 542-76
Sulston, J.E., Schierenberg, E., White, J.G. and Thomson, J.N. (1983) 'The Embryonic Cell Lineage of the Nematode *Caenorhabditis elegans*', *Developmental Biology, 100*, 64-119
Taylor, C.E. and Brown, D.J.F. (1981) 'Nematode-Virus Interactions', in B.M. Zuckerman and R.A. Rhode (eds), *Plant Parasitic Nematodes*, Vol. 3, Academic Press, New York and London
Thomas, R.J. (1974) 'The Role of Climate in the Epidemiology of Nematode Parasitism in Ruminants', in A.E.R. Taylor and R. Muller (eds), *The Effects of Meteorological Factors upon Parasites*, Blackwell, Oxford
Tobler, H., Smith, K.D. and Ursprung, H. (1972) 'Molecular Aspects of Chromatin Diminution in *Ascaris lumbricoides*', *Developmental Biology, 27*, 190-203
Triantaphyllou, A.C. (1971) 'Genetics and Cytology', in B.M. Zuckerman, W.F. Mai and R.A. Rhode (eds), *Plant Parasitic Nematodes*, Vol. 2, Academic Press, New York and London
Triantaphyllou, A.C. and Hirschmann, H. (1964) 'Reproduction in Plant and Soil Nematodes', *Annual Review of Phytopathology, 2*, 57-80
Tribe, H.T. (1980) 'Prospects for the Control of Plant-parasitic Nematodes', *Parasitology, 81*, 619-39
Vanfleteren, J.R. (1980) 'Nematodes as Nutritional Models', in B.M. Zuckerman (ed.), *Nematodes as Biological Models*, Vol. 2, Academic Press, New York and London
Veech, J.A. (1981) 'Plant Resistance to Nematodes', in B.M. Zuckerman and R.A. Rhode (eds), *Plant Parasitic Nematodes*, Vol. 3, Academic Press, New York and

London
Von Brand, T. (1973) *Biochemistry of Parasites*, Academic Press, New York and London
Vrain, T.C. (1978) 'Influence of Chilling and Freezing Temperatures on Infectivity of *Meloidogyne incognita* and *M. hapla*', *Journal of Nematology, 10*, 177-80
Vranken, G. and Heip, C. (1983) 'Calculation of the Intrinsic Rate of Natural Increase, r_m, with *Rhabditis marina* Bastian 1865 (Nematoda)', *Nematologica, 29*, 468-77
Wakelin, D. (1978) 'Immunity to Intestinal Parasites', *Nature, 273*, 617-20
Wakelin, D. (1984a) *Immunity to Parasites: How Animals Control Parasitic Infections*, Edward Arnold, London
Wakelin, D. (1984b) 'Evasion of the Immune Response: Survival within Low Responder Individuals of the Host Population', *Parasitology, 88*, 639-57
Wallace, H.R. (1963) *The Biology of Plant Parasitic Nematodes*, Edward Arnold, London
Wallace, H.R. (1968) 'Dynamics of Nematode Movement', *Annual Review of Phytopathology, 6*, 91-114
Wallace, H.R. (1973) *Nematode Ecology and Plant Disease*, Edward Arnold, London
Walsh, J.A., Shepherd, A.M. and Lee, D.L. (1983) 'The Distribution and Effect of Intracellular Rickettsia-like Micro-organisms Infecting Second-stage Juveniles of the Potato Cyst-nematode *Globodera rostochiensis*', *Journal of Zoology, London, 199*, 395-419
Ward, S. (1976) 'The Use of Mutants to Analyse the Sensory Nervous System of *Caenorhabditis elegans*', in N.A. Croll (ed.), *The Organisation of Nematodes*, Academic Press, New York and London
Ward, S. (1978) 'Nematode Chemotaxis and Chemoreceptors', in G.L. Hazelbauer (ed.), *Taxis and Behaviour: Elementary Sensory Systems in Biology*, Chapman & Hall, London
Ward, S. and Carrel, J.S. (1979) 'Fertilisation and Sperm Competition in the Nematode *Caenorhabditis elegans*', *Developmental Biology, 73*, 304-21
Ward, S., Thomson, J.W., White, J.G. and Brenner, S. (1975) 'Electronmicroscopical Reconstruction of the Anterior Sensory Anatomy of the Nematode *Caenorhabditis elegans*', *Journal of Comparative Neurology, 160*, 313-38
Ware, R.W., Clark, C., Crossland, K. and Russell, R.C. (1975) 'The Nerve Ring of the Nematode *Caenorhabditis elegans*', *Journal of Comparative Neurology, 162*, 71-110
Weinstein, P.P. (1952) 'Regulation of Water Balance as a Function of the Excretory System of the Filariform Larvae of *Nippostrongylus muris* and *Ancylostoma caninum*', *Experimental Parasitology, 1*, 363-76
Wharton, D.A. (1979a) 'Oogenesis and Egg-shell Formation in *Aspiculuris tetraptera* Schulz (Nematoda: Oxyuroidea)', *Parasitology, 78*, 131-43
Wharton, D.A. (1979b) 'The Structure and Formation of the Egg-shell of *Hammerschmidtiella diesingi* Hammerschmidt (Nematoda: Oxyuroidea)', *Parasitology, 79*, 1-12
Wharton, D.A. (1979c) 'The Structure and Formation of the Egg-shell of *Syphacia obvelata* Rudolphi (Nematoda: Oxyurida)', *Parasitology, 79*, 13-28
Wharton, D.A. (1980) 'Nematode Egg-shells', *Parasitology, 81*, 447-63
Wharton, D.A. (1981) 'The Effect of Temperature on the Behaviour of the Infective Larvae of *Trichostrongylus colubriformis*', *Parasitology, 82*, 269-79
Wharton, D.A. (1982a) 'Observations on the Coiled Posture of Trichostrongyle Infective Larvae Using a Freeze-substitution Method and Scanning Electron Microscopy', *International Journal for Parasitology, 12*, 335-43

Wharton, D.A. (1982b) 'The Structure of the Body Coverings of the Infective Larva of *Trichostrongylus colubriformis*', *Proceedings of the 5th International Congress of Parasitology*, Toronto

Wharton, D.A. (1982c) 'The Survival of Desiccation by the Free-living Stages of *Trichostrongylus colubriformis* (Nematoda: Trichostrongylidae)', *Parasitology, 84*, 455-62

Wharton, D.A. (1983) 'The Production and Functional Morphology of Helminth Egg-shells', *Parasitology, 86*, 85-97

Wharton, D.A. and Barrett, J. (1985) 'Ultrastructural Changes during Recovery from Anabiosis in the Plant Parasitic Nematode, *Ditylenchus*', *Tissue and Cell, 17*, 79-96

Wharton, D.A. and Jenkins, T. (1978) 'Structure and Chemistry of the Egg-shell of a Nematode (*Trichuris suis*)', *Tissue and Cell, 10*, 427-40

Wharton, D.A. and Sommerville, R.I. (1984) 'The Structure of the Excretory System of the Infective Larvae of *Haemonchus contortus*', *International Journal for Parasitology, 14*, 591-600

Wharton, D.A., Perry, R.N. and Beane, J. (1983) 'The Effect of Osmotic Stress on Behaviour and Water Content of Infective Larvae of *Trichostrongylus colubriformis*', *International Journal for Parasitology, 13*, 185-90

Wharton, D.A., Young, S.R. and Barrett, J. (1984) 'Cold Tolerance in Nematodes', *Journal of Comparative Physiology, B154*, 73-7

Wharton, D.A., Barrett, J. and Perry, R.N. (1985) 'Water Uptake and Morphological Changes During Recovery from Anabiosis in the Plant-parasitic Nematode, *Ditylenchus dipsaci*', *Journal of Zoology (London), 206*, 391-402

White, J.G., Southgate, E., Thomson, J.N. and Brenner, S. (1976) 'The Structure of the Ventral Nerve Cord of *Caenorhabditis elegans*', *Philosophical Transactions of the Royal Society of London, B275*, 327-48

Willett, J.D. (1980) 'Control Mechanisms in Nematodes' in B.M. Zuckerman (ed.), *Nematodes as Biological Models*, Vol. 1, Academic Press, New York and London

Wilson, P.A.G. (1976) 'Nematode Growth Patterns and the Moulting Cycle: the Population Growth Profile', *Journal of Zoology (London), 179*, 135-51

Wilson, P.A.G. (1982) 'Roundworm Juvenile Migrations in Mammals: the Pathways of Skin-penetrators Reconsidered', in E. Meerovitch (ed.), *Aspects of Parasitology*, McGill University, Montreal

Wisse, E. and Daems, W.T. (1968) 'Electron Microscope Observations on Second-stage Larvae of the Potato Root Eelworm *Heterodera rostochiensis*', *Journal of Ultrastructural Research, 24*, 210-31

Womersley, C. (1981) 'Biochemical and Physiological Aspects of Anhydrobiosis', *Comparative Biochemistry and Physiology, 70B*, 669-78

Woodhead-Galloway, J. (1980) *Collagen: the Anatomy of a Protein*, Edward Arnold, London

Wright, D.J. (1981) 'Nematocides: Modes of Action and New Approaches to Chemical Control', in B.M. Zuckerman and R.A. Rhode (eds), *Plant Parasitic Nematodes*, Vol. 3, Academic Press, New York and London

Wright, D.J. and Newall, D.R. (1976) 'Nitrogen Excretion, Osmotic and Ionic Regulation in Nematodes', in N.A. Croll (ed.), *The Organisation of Nematodes*, Academic Press, New York and London

Wright, D.J. and Newall, D.R. (1980) 'Osmotic and Ionic Regulation in Nematodes', in B.M. Zuckerman (ed.), *Nematodes as Biological Models*, Vol. 2, Academic Press, New York and London

Wright, E.J. and Sommerville, R.I. (1977) 'Movement of a Non-flagellate Spermatozoon: a Study of the Male Gamete of *Nematospiroides dubius* (Nematoda)', *International Journal for Parasitology, 7*, 353-9

Wright, K.A. (1976) 'Functional Organisation of the Nematode's Head', in N.A. Croll (ed.), *The Organisation of Nematodes*, Academic Press, New York and London

Wright, K.A. (1980) 'Nematode Sense Organs', in B.M. Zuckerman (ed.), *Nematodes as Biological Models*, Vol. 2, Academic Press, New York and London

Wright, K.A. (1983) 'Nematode Chemosensillae: Form and Function', *Journal of Nematology, 15*, 151-8

Wu, Y-J. and Foor, W.E. (1983) 'Ultrastructure and Function of Oviduct-Uterine Junction in *Ascaris suum* (Nematoda)', *Journal of Parasitology, 69*, 121-8

Wyss, V. (1981) 'Ectoparasitic Root Nematodes: Feeding Behaviour and Plant Cell Responses', in B.M. Zuckerman and R.A. Rhode (eds), *Plant Parasitic Nematodes*, Vol. 3, Academic Press, New York and London

Yuksel, H.J. (1960) 'Observations on the Life Cycle of *Ditylenchus dipsaci* on Onion Seedlings', *Nematologica, 5*, 289-96

Zengel, J.M. and Epstein, H.F. (1980) 'Muscle Development in *Caenorhabditis elegans*: a Molecular Genetic Approach', in B.M. Zuckerman (ed.), *Nematodes as Biological Models*, Vol. 1, Academic Press, New York and London

Zuckerman, B.M. (ed.) (1980) *Nematodes as Biological Models*, Vols 1 and 2, Academic Press, New York and London

Zuckerman, B.M. and Jansson, H.B. (1984) 'Nematode Chemotaxis and Possible Mechanisms of Host/Prey Recognition', *Annual Review of Phytopathology, 22*, 95-113

Zuckerman, B.M., Kahane, I. and Himmelhoch, S. (1979) '*Caenorhabditis briggsae* and *C. elegans*: Partial Characterisation of Cuticle Surface Carbohydrates', *Experimental Parasitology, 47*, 419-24

INDEX